Ayat Khaled Sagheer
Kareema Abed Alkadim Makrib AlKhafaji

Modeling for the Lifetime Data with Applications

Ayat Khaled Sagheer
Kareema Abed Alkadim Makrib AlKhafaji

Modeling for the Lifetime Data with Applications

Modeling for the Lifetime Data withRayleigh-Exponentiated Exponential Distribution (R-EE)

Noor Publishing

Cover image: www.ingimage.com

Publisher:
Noor Publishing
is a trademark of
Dodo Books Indian Ocean Ltd. and OmniScriptum S.R.L publishing group

120 High Road, East Finchley, London, N2 9ED, United Kingdom
Str. Armeneasca 28/1, office 1, Chisinau MD-2012, Republic of Moldova, Europe
Managing Directors: Ieva Konstantinova, Victoria Ursu
info@omniscriptum.com

Printed at: see last page
ISBN: 978-620-0-04902-5

Modeling for the Lifetime Data with Applications

By

Ayat Khaled Sagheer

Prof. Dr. Kareema Abed Al- kadim Makrib Al- Khafaji

List of Contents

List of Abbreviations

Symbol	Description
F(.)	The Cumulative Distribution Function (cdf).
f(.)	The Probability Density Function (pdf).
$\bar{F}(.)$	The Reliability (Survivor) Function.
h(.)	The Hazard Function.
H(.)	The Cumulative Hazard Function.
r (.)	The reverse Hazard Function
μ'_r	The rth Moment about the Origin
μ_r	The rth Moment about the Mean
μ	The Mean
$\Gamma(.)$	Gamma Function
$M_X(.)$	The Moment Generating Function
$I_R(\delta)$	Rényi Entropy

List of Symbols

Abbreviations	Description
CS	Coefficient of Skewness.
Ck	Coefficient of Kurtosis .
CV	Coefficient of Variation .
AIC	Akaike Information Criterion.
HQIC	Hannan-Quinn Information Criterion .
BIC	Bayesian Information Criterion.
CAIC	Corrected Akaike Information Criterion.
MGF	The Moment Generating Function

List of Tables

List of Figures

Abstract

In our life there are many distribution that can be used to model life time data. However there are many important problem in which the real data cannot fit any of the previous life time distributions .

Wherefore this works deals with new two flexible distributions, which are (the Weibull – Exponentiated Exponential and the Rayleigh_ Exponentiated Exponential) , by finding the cumulative probability and probability density functions of the two distributions and discussion some of statistical and mathematical properties of each moments , variances , the coefficient of skewness , the generating function of moments , entropy , and ordered statistics .

The parameters of each of these distribution have estimated using the maximum likelihood , moment method. The useful , of their distribution has explained by application of two real data sets .

The proposed distribution are better than (Weibull _Exponentiated Exponential ,Exponential – Weibull , odd Generalized Exponential , Flexible_Weibull ; Weibull Rayleigh , Rayleigh _Rayleigh , extended odd Weibull Rayleigh) through the comparison between them.

All computation and plot have done by using program Matlab (R2012b).

Introduction

In statistical analysis , one of the main tasks of statistics is to find efficient statistical modeling for natural life events in the form of known probability distributions , probability distributions are used to model natural life phenomena such as the exponential distribution, Weibull distribution, normal distribution, gamma distribution and other distributions play an important role in real life areas such as reliability, survival analysis, ecology, medicine and social sciences.

Many probability distributions have been derived because the natural life phenomena are complex and very hard to model by traditional distributions. However, known probability distributions remain unable to represent data for some natural phenomena accurately.

This lead to the expansion and modification of generalized probability distributions. The addition of some new parameters to the known probability distributions improved the quality of suitability for the natural phenomena data and higher accuracy of describing the shape of the distribution's tail. Our goal is to develop these distributions through generalization to become more flexible or more compatible with the data set by generalizing common life distributions such as Weibull , exponential, gamma, etc.

In statistical literature , exponential distribution was the first widely used lifetime model in many fields of the real life such as medicine, ecology and reliability. There has been an increased interest in developing Exponentiated families of distributions by introducing one or more additional shape parameter to the base distribution. The Exponentiated families of distributions has attracted the attention of statisticians, mathematicians scientists , engineers , economists , and demographers .

One reason might be the computational and analytical facilities available in programming software's like R (packages) , and Matlab. The second reason is the properties of E distributions that can easily be explored. Thirdly, the shape parameter helpful in improving the goodness-of-fit of the proposed EF. Fourthly, EF has the ability to fit skewed data better than other distributions [23] .

Exponentiated family(EF) [23] of distributions established since the first half of the nineteenth century when Gompertz (1825) and Verhulst (1838, 1845, 1847) used the cdf .

Exponentiated Exponential distributions have increasing as well as decreasing hazard function depending on the value of the shape parameter, also they have a constant hazard function when the shape parameter is equal to one. (Gupta and Kundu, 1999)[13] ,[14] .

The Weibull distribution is one of the most popular and widely used model for failure time in life-testing and reliability theory. However, a drawback of this distribution as far as lifetime analysis is concerned is the monotonic behaviour of its hazard date function .

In real life applications, empirical hazard rate curves often exhibit non-monotonic shapes such as a bathtub, upside-down bathtub (unimodal) and others. So, there is a gen uine desire to search for some generalizations or modifications of the Weibull distribution that can provide more flexibility in lifetime modeling.

In the recent literature , four Weibull based generators have appeared , namely : Exponentiated Weibull-G (EW-G) [16] , Exponentiated extended-G [9].

Most of the Rayleigh probability distribution extensions have been derived because of their great importance in describing many natural life phenomena.

The Rayleigh probability density function, attributed to Lord Rayleigh (1842–1919), is concerned with describing skewed data see Rayleigh [27] . Mahmoud and Ghazal [22] discussed the estimation of the Exponentiated Rayleigh parameters based on type II censored data .

Here we study a new model with four- parameters continuous distribution called the Weibull-Exponentiated Exponential distribution $(\mathbf{W-EE})$ when the cumulative distribution function for the as Exponentiated Exponential (EE) .

By using multiply $\left(\boldsymbol{x^b}\right)$ in cumulative distribution equation to Exponentiated Exponential distribution by cumulative distribution equation to Weibull distribution then we get the cdf of (W − EE) distribution .

And by using the same is true for Rayleigh distribution by equation cdf the Exponentiated Exponential distribution then we get the cdf $(R-EE)$ distribution. Then we find the corresponding pdfs of these distributions by driving the cdfs of these distribution.

All of the figures and the values in tables are found by using MATLAB **(R2012b)** software .

This works consists of four chapters as follows:

Chapter One consists of definitions and principal concepts like some properties of the Weibull , Exponential and Rayleigh distributions , functions related to reliability , definition of coefficients , of skewness , kurtosis and variation , useful functions.

Chapter Two We present and study a new distribution called the W-EE distribution, the cumulative distribution function cdf (F(x)) , the probability density function pdf (f(x)) , functions of reliability like the reliability function $\bar{F}(x)$, the hazard function h(x) , the cumulative hazard function chf (H(x)), and the reverse hazard function rhf (r (x)) of this distribution are proposed , we also drive some properties of this distribution such as , the mean , r^{th} moment about the origin, r^{th} moment about the mean . Coefficient of skewness, kurtosis and variation , the moment generating function , order statistics , and Rényi entropy are defined . The four parameters of the W-EE distribution are estimated by using the maximum likelihood estimation method . We illustrate the importance of the new family by means of two applications to a real data sets.

Chapter Three contains and studies another new distribution called the R-EE distribution, the cdf $(F(x))$, pdf $\big(f(x)\big)$, and study as the properties of the distribution in chapter two.

Chapter Four Contains conclusions and future work are introduced .

Publication

1 . Ayat Khaled Sagheer , and Kareema Abed AL-Kadim. "Modeling Lifetime Data With Weibull-Exponentiated Exponential Distribution (W-EE) ." Journal of Physics: Conference Series. IOP Publishing. Accepted Publication.

2. Ayat Khaled Sagheer , and Kareema Abed AL-Kadim. "Modeling Lifetime Data With rayleigh -Exponentiated Exponential Distribution (R-EE) ." AIP Conference proceedings scopus Indexed . Accepted Publication.

Chapter One

Definitions and Principal Concepts

1.1 Properties of the Used Distributions

In this section, we introduce properties of the Weibull , Exponentiated Exponential and Rayleigh distributions.

1.1.1 Properties of the Weibull Distribution [3]

The Weibull distribution is a widely known distribution due to its utility in modeling lifetime data. A continuous random variable X is said to have a Weibull distribution (W.D.), if it has the following probability density function:

$$f(x;a,b) = \begin{cases} abx^{b-1}e^{-ax^b} & , \quad x \geq 0 \\ 0 & , \quad x < 0 \end{cases} \tag{1.1}$$

And the cdf is defined by :

$$F(x;a,b) = 1 - e^{-ax^b}, \quad x \geq 0 \tag{1.2}$$

Table 1.1: Properties of the Weibull Distribution.

Mean	$\left(\frac{1}{a}\right)^{\frac{1}{b}}\Gamma(1+b^{-1})$
Variance	$\left(\frac{1}{a}\right)^{\frac{2}{b}}[\Gamma(1+2b^{-1}) - \Gamma^2(1+b^{-1})]$
Mode	$\begin{cases} \left(\frac{1}{a}\right)^{\frac{1}{b}}\left(\frac{b-1}{b}\right)^{\frac{1}{b}} & b \geq 1 \\ 0 & b < 1 \end{cases}$
CS	$\frac{\Gamma(1+\frac{3}{b})(\frac{1}{a})^{\frac{3}{b}} - 3\mu\sigma^2 - \mu^3}{\sigma^3}$
MGF	$\frac{\Gamma(\alpha+1)\Gamma(1-t)}{\Gamma(\alpha-t+1)}$

1.1.2 Properties of the Exponential Distribution [19]

The Exponential distribution is commonly used to model waiting times between occurrence of rare events and lifetime of electrical or mechanical devices. A continuous random variable X is said to have an Exponential Distribution with parameter ($\lambda > 0$), if it has the following probability density function:

$$f(x;\lambda) = \begin{cases} \lambda e^{-\lambda x} & x \geq 0 \\ 0 & x < 0 \end{cases} , \qquad (1.3)$$

and the cdf is defined by

$$F(x;\lambda) = 1 - e^{-\lambda x} \qquad x \geq 0 \qquad (1.4)$$

The Exponentiated Exponential (EE) [12] distribution is defined

$$F(x;\lambda,\alpha) = \left[1 - e^{-\lambda x}\right]^{\alpha} \qquad , x \geq 0, \alpha, \lambda \geq 0 \qquad (1.5)$$

Exponential distribution has the following properties:

Table 1.2: properties of the exponential distribution.

Mean	$\frac{1}{\lambda}$
Variance	$\frac{1}{\lambda^2}$
Mode	0
CS	2
MGF	$\frac{\lambda}{\lambda - t}, t < \lambda$

1.1.3 Properties of the Rayleigh Distribution [11]

The Rayleigh distribution is often used where two orthogonal components have an absolute value, for example, wind velocity and direction may be combined to yield a wind speed, or real and imaginary components may have absolute values that are Rayleigh distributed .

A random variable X is said to have the Rayleigh distribution (RD) with parameter θ if its probability density function is given by

$$f(X,\lambda) = \frac{x}{\lambda^2} e^{\frac{-x^2}{2\lambda^2}}, \ x > 0, \lambda > 0 \qquad (1.6)$$

While the cumulative distribution function is given by

$$F(X,\lambda) = \left[1 - e^{\frac{-x^2}{2\lambda^2}}\right], x > 0, \lambda > 0 \qquad (1.7)$$

where λ denote the scale parameter.

Table 1.3: properties of the Rayleigh Distribution.

Mean	$\sqrt{\frac{\pi}{2}}\lambda$
Variance	$\frac{4-\pi}{2}\lambda^2$
Mode	λ
CS	$\frac{2\sqrt{\pi}(\pi-3)}{(4-\pi)^{\frac{3}{2}}}$
MGF	$1 + \lambda t e^{\lambda^2 \frac{t^2}{2}} \sqrt{\frac{\pi}{2}} (\text{erf}^* \left(\frac{\lambda t}{\sqrt{2}}\right) + 1)$ erf*= error function

1.2 Functions Related to Reliability [28] , [5]

Let X be a continuous random variable of distribution that has a parameter vector $\theta = \{\theta_1, \theta_2, \theta_3, \dots, \theta_n\}$, and let $f(x;\theta)$ be the pdf of X , then the cdf of X is given by :

$$F(x;\ \theta) = P(X \le x) = \int_{-\infty}^{x} f(t;\theta)dt, \tag{1.8}$$

The **Reliability Function** works with devices (the **Survival Function** works with persons) taken from [28] is defined as the probability that a system will operate satisfactorily for a given period of time and denoted by $\overline{F}(x)$ where

$$\overline{F}(x,\theta) = P(X > x) = 1 - F(x;\ \theta). \tag{1.9}$$

Therefore from equations (1.7) and (1.8), we get:

$$F(x;\ \theta) + \overline{F}(x,\theta) = 1\,. \tag{1.10}$$

The **Hazard Function** $h(x;\ \theta)$ [28] for any continuous distribution with pdf $f(x;\theta)$ and cdf $F(x;\ \theta)$ is defined as follows :

$$\boldsymbol{h}(x;\ \theta) = \frac{f(x;\theta)}{\overline{F}(x,\theta)}, \tag{1.11}$$

The **Reverse Hazard Function** $r(x;\ \theta)$ [5] is defined as follows :

$$\boldsymbol{r}(x;\ \theta) = \frac{f(x,0)}{F(x;\ \theta)}, \tag{1.12}$$

And the **Cumulative Hazard Function** [4] is defined as follows :

$$H(x;\ \theta) = \int_0^x h(t;\ \theta)dt = \int_0^x \frac{f(t;\theta)}{R(t;\ \theta)}dt = -\ln(R(x;\ \theta)) \tag{1.13}$$

1.3 Definition of Coefficients [8],[24],[30]

In this section, we will define the coefficient of skewedness, kurtosis, and variation.

1.3.1 Coefficient of Skewness [24]

The coefficient of Skewness is the degree of symmetry of the distribution's curve , it is denoted by $\boldsymbol{CS}$, the skewness is zero for the symmetric distribution when $(\textbf{mean}(\mu) = \boldsymbol{median}(\mu_e) = \boldsymbol{mode}(\mu_o))$, positive for the skewed to the right distribution when ($\mu > \mu_e > \mu_o$) and negative if the distribution is skewed to the left when ($\mu < \mu_e < \mu_o$) , and it is defined by

$$CS = \frac{E(X-\mu)^3}{\left(\sqrt{var(x)}\right)^3}, \tag{1.14}$$

Where $E(X-\mu)^3$ is center moment about the mean when $r = 3$, and $\sqrt{var(x)}$ is the standard deviation of distribution .

1.3.2 Coefficient of Kurtosis[30]

It measures the flatness of the top of the curve , the kurtosis equals 3 for the normal distribution , positive for the more tall and slim curves than the normal one in the neighborhood of the mode when (**CK > 3**) (in this case it is said to be leptokurtic) and it is negative for platykurtic distribution when (**CK< 3**) (it means that the distribution is flatter than the normal distribution) and it is defined by:

$$CK = \frac{E(X-\mu)^4}{(var(X))^2}, \tag{1.15}$$

Where $\boldsymbol{E(X-\mu)^4}$ the center moment about the mean when $r = 4$, $\boldsymbol{var(X)}$ the variance of distribution .

1.3.3 Coefficient of Variation [8]

The coefficient of variation is a dimensionless number which quantifies the degree of variability relative to the mean. The population coefficient of variation is defined as:

$$CV = \frac{\sqrt{var(X)}}{E(X)} , \tag{1.16}$$

Where $\sqrt{\boldsymbol{var(X)}}$ is the standard deviation of distribution and $\boldsymbol{E(X)}$ is the expection of distribution .

1.4 Useful Functions [17],[18],[33]

In this section , we will define of the Gamma Function , and Binomial Theorem Formula .

1.4.1. Gamma Function [17]

In mathematics , the gamma function (represented by Γ, the capital letter gamma from the Greek alphabet) is one commonly used extension of the factorial function to complex numbers.

The gamma function is defined for all complex numbers except the non-positive integers For any positive integer n [18]

$$\Gamma(n) = (n-1)! \tag{1.17}$$

The gamma function [17] is given by:

$$\Gamma(x) = \int_0^\infty t^{x-1} e^{-t} dt , \qquad x > 0 \tag{1.18}$$

We can write the gamma function as follows:

$$\frac{\Gamma(x)}{a^x} = \int_0^\infty t^{x-1} e^{-at} dt , \qquad x > 0 \tag{1.19}$$

The first and the second derivative of the gamma function [17] , where x>0 are given by :

$$\Gamma'(x) = \int_0^\infty t^{x-1} (\log t)\, e^{-t} dt \tag{1.20}$$

and

$$\Gamma''(x) = \int_0^\infty t^{x-1} (logt)^2\, e^{-t} dt , \qquad x > 0. \tag{1.21}$$

The connection between the beta and the gamma function is defined in [18] as follows:

$$B(a, b) = \frac{\Gamma(a)\Gamma(b)}{\Gamma(\mathrm{a} + \mathrm{b})} \tag{1.22}$$

1.4.2 Binomial Theorem Formula [33]

Based on the binomial properties, the binomial theorem states that the following **binomial formula** is valid for all positive integer values of *n*:

$$(a + b)^n = \sum_{i=0}^{n} \binom{n}{i} a^i b^{n-i} ,$$

where n is positive integer number .

$$(a-b)^n = \sum_{i=0}^{n} (-1)^i \binom{n}{i} a^i b^{n-i} \,, \tag{1.22}$$

Newton (1676) showed the formula also holds for negative integers $-n$, [33]

$$(a+b)^{-n} = \sum_{i=0}^{\infty} \binom{-n}{i} a^i b^{-n-i} \,, \tag{1.23}$$

Chapter Two

Modeling for the Lifetime Data with Weibull-Exponentiated Exponential Distribution (W − EE)

2.1 Introduction

In this chapter a new distribution , inducing a new four- parameters continuous distribution called the Weibull-Exponentiated Exponential distribution (**W − EE**) for modeling life time data that created in a new way by using two life distributions .

The Weibull Exponentiated Exponential distribution has been used to model four lifetime data sets and the results show that the proposed generalization performs better than the other known extensions of the Weibull distribution considered for the study.

We studied the four parameters Weibull - Exponentiated Exponential distribution $W - EE\ (a, b, \alpha, \lambda)$ distribution .

2.2 The CDF and PDF of $(W - EE)$ Distribution

We can find the cdf and pdf of $(W - EE)$ distribution by using the cdf of the weibull distribution that given by :

$$F(X; a, b) = \left[1 - e^{-ax^b}\right], x > 0, a, b > 0 \qquad (2.1)$$

And the cdf of the Exponentiated exponential distribution that given by :

$$F(X; \alpha, \lambda) = [1 - e^{-\alpha x}]^{\lambda}, \qquad x > 0, \alpha > 0, \lambda > 0 \qquad (2.2)$$

as the follows :

We multiply $(\boldsymbol{x^b})$ in eq. (2.1) by eq. (2.2) then we get the cdf of $(W - EE)$ distribution as the following equation:

$$F(X; a, b, \alpha, \lambda) = 1 - e^{-ax^b[1-e^{-\alpha x}]^{\lambda}}, x, \alpha, \lambda, a, b > 0 \qquad (2.3)$$

And the pdf corresponding to equation (2.3) is given by :

$$f(X;a,b,\alpha,\lambda)$$

$$= ae^{-ax^b[1-e^{-\alpha x}]^\lambda}\left[x^b\alpha\lambda e^{-\alpha x}[1-e^{-\alpha x}]^{\lambda-1} + bx^{b-1}[1-e^{-\alpha x}]^\lambda\right] ,$$

$x > 0$, $\alpha, \lambda, a, b > 0$, (2.4)

$$f(X;a,b,\alpha,\lambda) = ax^b e^{-ax^b[1-e^{-\alpha x}]^\lambda}[1-e^{-\alpha x}]^\lambda\left[\frac{\alpha\lambda}{e^{\alpha x}-1}+\frac{b}{x}\right]$$

$x > 0$, $\alpha, \lambda, a, b > 0$, (2.5)

Where b *and* λ are shape parameters and a *and* α are scale parameters.

Āow we prove it is pdf

1. $f(X;a,b,\alpha,\lambda) \geq 0$
2. $\int_0^\infty f(X;a,b,\alpha,\lambda)dx = 1$

That is

1. $f(X;a,b,\alpha,\lambda) \geq 0$

In (EE) distribution $x \in [0,\infty)$ then in W-EE distribution $x \in [0,\infty)$

When $x \to 0$, then $f(X;\alpha,\lambda,a,b) = 0$ and as $x \to \infty$, then $f(x;\alpha,\lambda,a,b) = 0$

$$\therefore\ f(x;\alpha,\lambda,a,b) \geq 0$$

2. $\int_0^\infty f(X;a,b,\alpha,\lambda) =$

$$= \int_0^\infty ae^{-ax^b[1-e^{-\alpha x}]^\lambda}\left[x^b\alpha\lambda[1-e^{-\alpha x}]^{\lambda-1}e^{-\alpha x} + bx^{b-1}[1-e^{-\alpha x}]^\lambda\right]$$

$$= \left[-e^{-ax^b[1-e^{-\alpha x}]^\lambda}\right]\Big|_0^\infty = 1 \qquad (2.6)$$

The plots of the cumulative distribution function $F(X;a,b,\alpha,\lambda)$ are given by the following figures

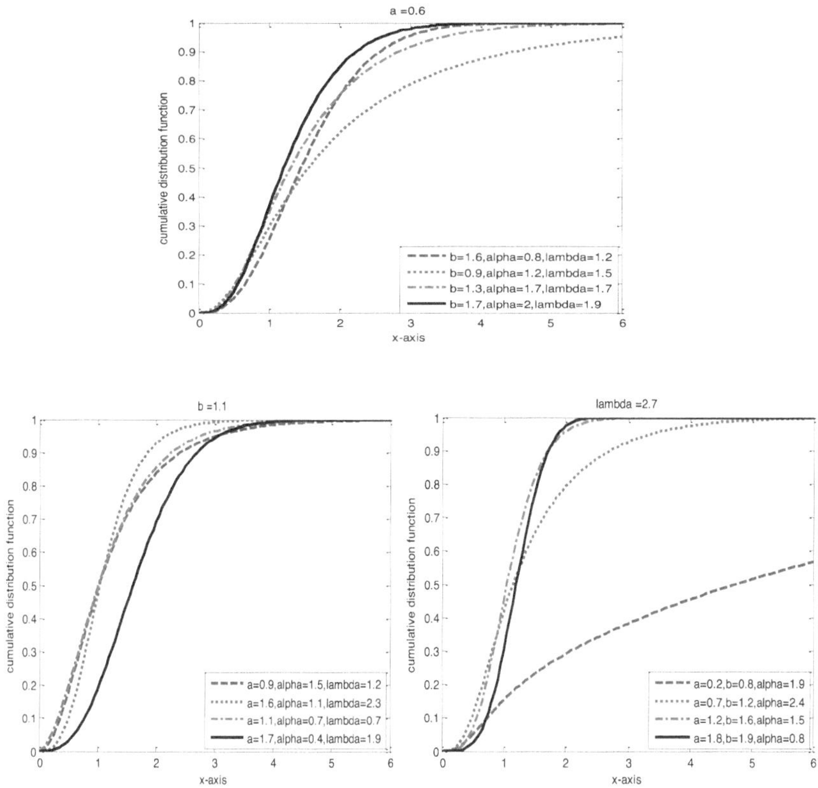

Figure (2.1) : The Cdf of $W-EE$ distribution indicate that the cdf of the $W-EE$ is non-decreasing with increasing x and the parameters α , λ , a , b

The plots of the probability density function $f(X; a, b, \alpha, \lambda)$ are given by the following figures .

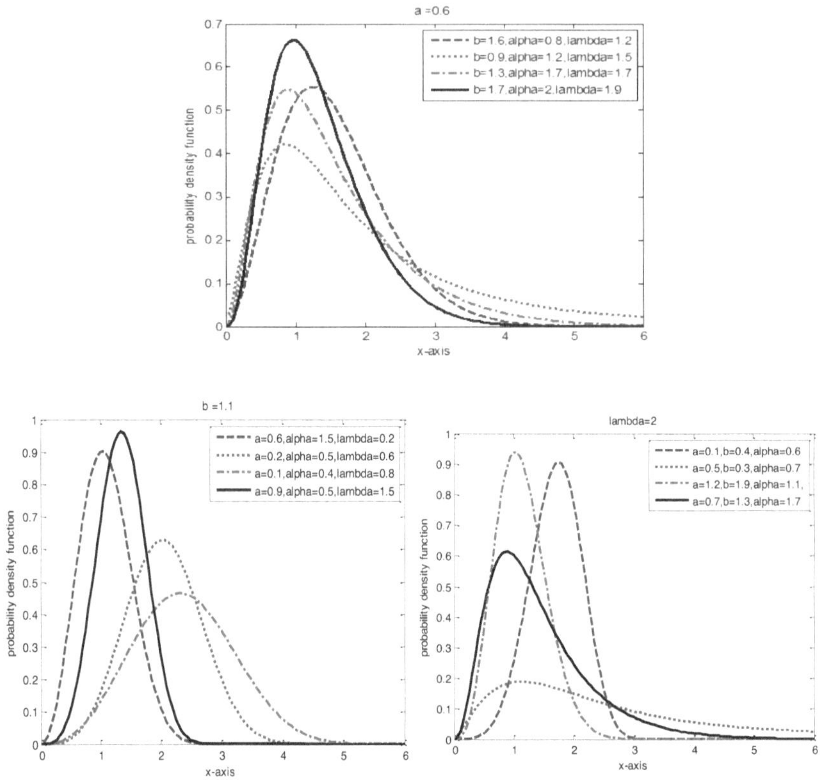

Figure (2.2) : The Pdf of $W-EE$ distribution with the parameters , indicate that the $W-EE$ family generates various shapes such as symmetrical , right skewed and reversed-J

2.3 The Limit of the Pdf and Cdf of W-EE

The limit of the pdf is given as follows :

$$\lim_{x\to 0} f(X;a,b,\alpha,\lambda) = a\lim_{x\to 0}\left[e^{-ax^b[1-e^{-\alpha x}]^\lambda}\left[\left[x^b\alpha\lambda[1-e^{-\alpha x}]^{\lambda-1}\right]e^{-\alpha x} + bx^{b-1}[1-e^{-\alpha x}]^\lambda\right]\right] = 0 \quad (2.7)$$

$$\lim_{x\to\infty} f(X;a,b,\alpha,\lambda) = a\lim_{x\to\infty}\left[e^{-ax^b[1-e^{-\alpha x}]^\lambda}\left[\left[x^b\alpha\lambda[1-e^{-\alpha x}]^{\lambda-1}\right]e^{-\alpha x} + bx^{b-1}[1-e^{-\alpha x}]^\lambda\right]\right] = 0 \quad (2.8)$$

This is the pdf approaches to 0 as x approach to 0 or ∞ and that is shown Figure (2.2) .

The limit of cdf is given as follows :

$$\lim_{x\to 0} F(X;a,b,\alpha,\lambda) = \lim_{x\to 0}\left[1-e^{-ax^b[1-e^{-\alpha x}]^\lambda}\right] = 0 \quad (2.9)$$

$$\lim_{x\to\infty} F(X;a,b,\alpha,\lambda) = \lim_{x\to\infty}\left[1-e^{-ax^b[1-e^{-\alpha x}]^\lambda}\right] = 1 \quad (2.10)$$

This is the cdf approaches to 0 as x approach to 0 and ∞ and the cdf approaches to 1 as x approach to ∞ and that is shown figure (1) .

2.4 Reliability Function

In this section , we introduce the reliability (survival) function $\bar{F}(X)$, hazard function $h(X)$ and cumulative hazard rate function $H(X)$ of $\boldsymbol{X \sim W-EE(a,b,\alpha,\lambda)}$.

2.4.1 The Reliability (Survivor) Function

The reliability (survival) function describes the likelihood that a system or an individual will not fail after a given time.

It tells us about the probability of success or survival of a given product or component.

The reliability (survival) function of $\boldsymbol{X \sim W - EE(a, b, \alpha, \lambda)}$ is defined as follows.

$$\bar{F}(X; \alpha, \lambda, a, b) = 1 - F(X; \alpha, \lambda, a, b)$$

$$= 1 - \left[1 - e^{-ax^b[1-e^{-\alpha x}]^\lambda}\right] = \left[e^{-ax^b[1-e^{-\alpha x}]^\lambda}\right] \tag{2.11}$$

The plots of the reliability function $\bar{F}(x; a, b, \alpha, \lambda)$ are given by the following figures .

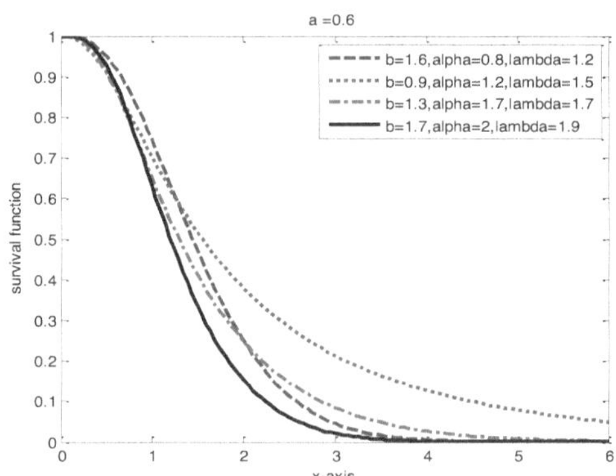

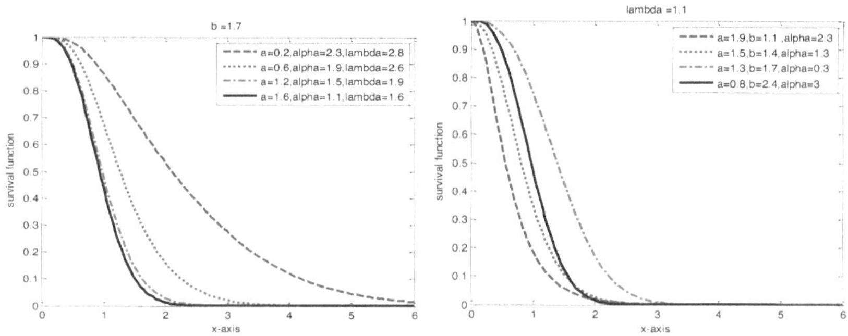

Figure (2.3) : The $\bar{F}(X)$ of $W - EE$ distribution , indicate that the $\bar{F}(X)$ of $W - EE$ is decrease function with decreasing x and the parameters α , λ , a , b.

2.4.2 Hazard Function

The hazard function , of $\boldsymbol{X \sim W - EE(a, b, \alpha, \lambda)}$ is defined as follows :

$$h(X; \alpha, \lambda, a, b) = \frac{f(X; a, b, \alpha, \lambda)}{\bar{F}(X; a, b, \alpha, \lambda)}$$

$$= \frac{a e^{-a x^b [1-e^{-\alpha x}]^\lambda} \left[x^b \alpha \lambda [1 - e^{-\alpha x}]^{\lambda - 1} e^{-\alpha x} + b x^{b-1} [1 - e^{-\alpha x}]^\lambda \right]}{\left[e^{-a x^b [1-e^{-\alpha x}]^\lambda} \right]} \tag{2.12}$$

The plots of the hazard function $\bar{F}(X; a, b, \alpha, \lambda)$ are given by the following figures.

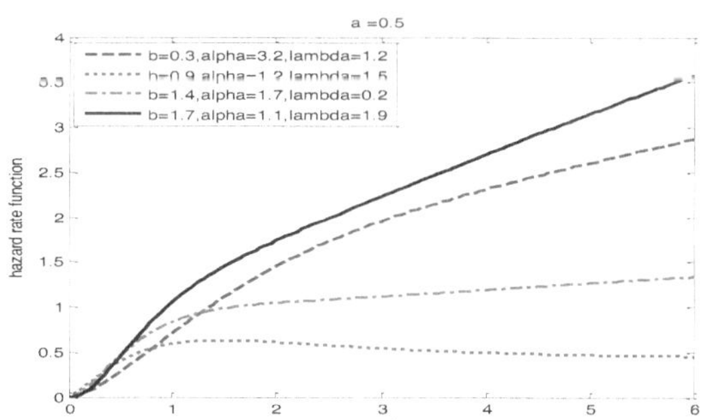

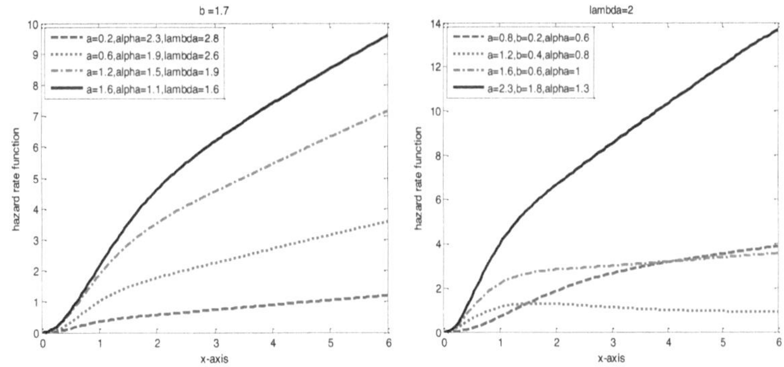

Figure (2.4) : The hazard function of W-EE distribution , indicate that the $h(X)$ of $W-EE$ is non-decreasing function

2.4.3 The Reverse Hazard Function [5]

The reverse hazard function of $\boldsymbol{X \sim W-EE}(a,b,\alpha,\lambda)$ is defined as follows

$$r(X,\alpha,\lambda,a,b) = \frac{f(X;a,b,\alpha,\lambda)}{F(X;a,b,\alpha,\lambda)}$$

$$= \frac{ae^{-ax^b[1-e^{-\alpha x}]^\lambda}\left[x^b\alpha\lambda[1-e^{-\alpha x}]^{\lambda-1}e^{-\alpha x} + bx^{b-1}[1-e^{-\alpha x}]^\lambda\right]}{\left[1-e^{-ax^b[1-e^{-\alpha x}]^\lambda}\right]} \quad (2.13)$$

The plots of the reverse hazard function of $W-EE$ are given by the following figures .

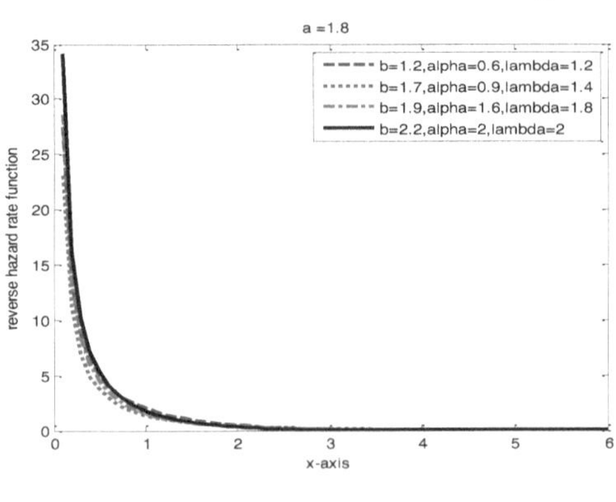

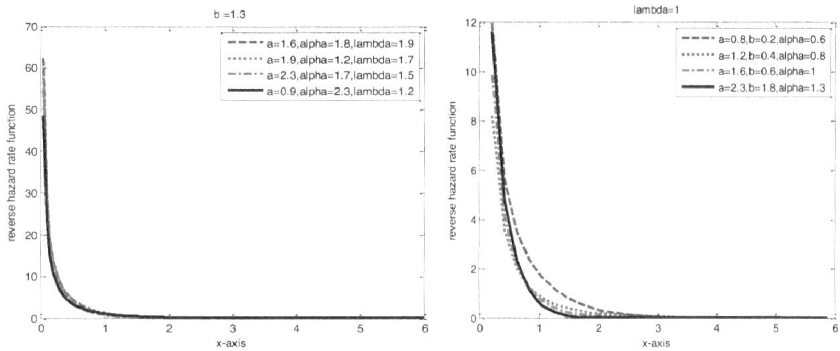

Figure (2.5) : The reverse hazard function of W-EE distribution , indicate that the $r(x)$ of $W-EE$ is non-decreasing function

2.4.4 The Cumulative Hazard Function [4]

The cumulative hazard function of $\boldsymbol{X \sim W-EE(a,b,\alpha,\lambda)}$ is defined as follows

$$H(X;a,b,\alpha,\lambda) = -\ln[1-F(X,a,b,\alpha,\lambda)]$$

$$= -\ln\left[1-\left[1-e^{-ax^b[1-e^{-\alpha x}]^\lambda}\right]\right]$$

$$= ax^b[1-e^{-\alpha x}]^\lambda \qquad (2.14)$$

The plots of the cumulative hazard function of $W-EE$ are given by the following figures.

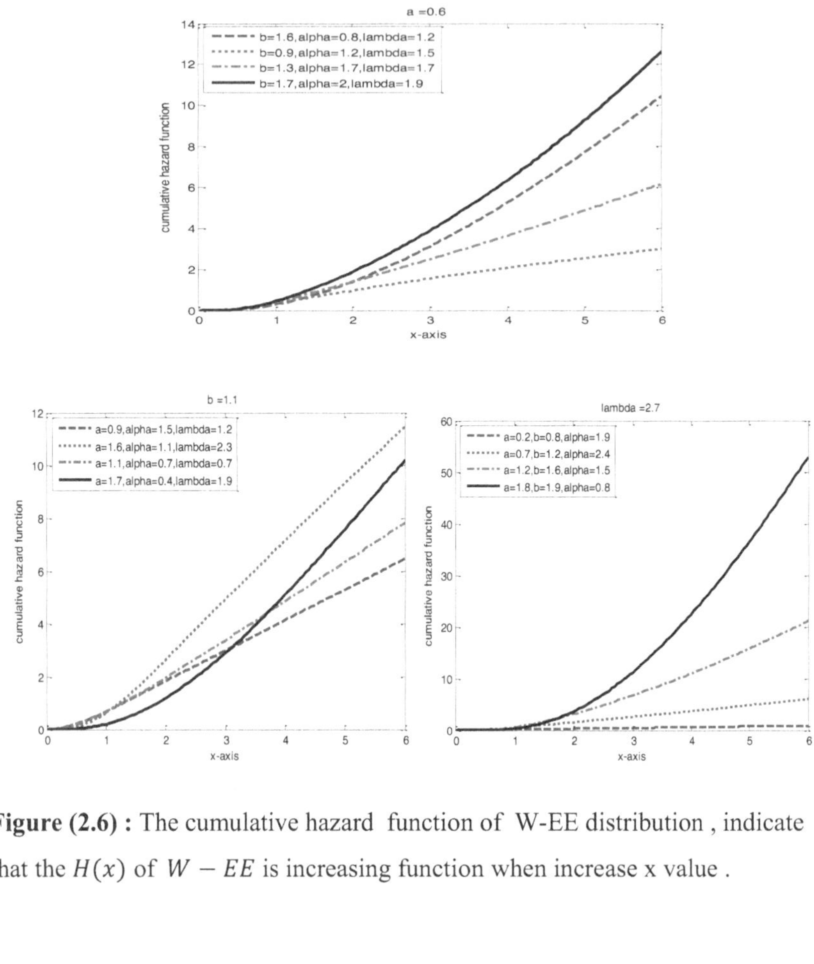

Figure (2.6) : The cumulative hazard function of W-EE distribution , indicate that the $H(x)$ of $W-EE$ is increasing function when increase x value .

2.5 The Moments and Coefficients of Skewness , Kurtosis and Variation

In this section , we introduce the r^{th}moment about the origin , r^{th} moment about the mean , coefficient of skewness , kurtosis and variation for the $W-EE$ distribution .

2.5.1 Theorem

Let $X \sim W-EE(a,b,\alpha,\lambda)$ the rth moment about the origin , mean respectively are as:

a) $\acute{\mu}_r = E(X^r) = a\alpha\lambda \sum_{j=0}^{\infty}\sum_{n=0}^{\infty} \frac{(-1)^{j+n}a^j}{j!}\binom{\lambda(j+1)-1}{n}\frac{\Gamma(r+b(j+1)+1)}{(\alpha n+\alpha)^{r+b(j+1)+1}} +$

$ab\sum_{j=0}^{\infty}\sum_{m=0}^{\infty} \frac{(-1)^{j+m}a^j}{j!}\binom{\lambda(j+1)}{m}\frac{\Gamma(r+b(j+1))}{(\alpha m)^{r+b(j+1)}}, r \in Z^+$

(2.15)

b) $\mu_r = E[(x-\mu)^r]$

$$= a\alpha\lambda \sum_{j=0}^{\infty}\sum_{n=0}^{\infty}\sum_{s=0}^{r}\binom{r}{s}\frac{(-1)^{j+n+r-s}a^j\mu^{r-s}}{j!}\binom{\lambda(j+1)-1}{n}\frac{\Gamma(s+b(j+1)+1)}{(an+\alpha)^{s+b(j+1)+1}}$$

$$+ ab\sum_{j=0}^{\infty}\sum_{m=0}^{\infty}\sum_{s=0}^{r}\binom{r}{s}\frac{(-1)^{j+m+r-s}a^j\mu^{r-s}}{j!}\binom{\lambda(j+1)}{m}\frac{\Gamma\big(s+b(j+1)\big)}{(am)^{s+b(j+1)}},$$

$r \in Z^+$

(2.16)

Proof a):

$$\acute{\mu}_r = a\int_0^{\infty} x^r e^{-ax^b[1-e^{-\alpha x}]^\lambda}\Big[x^b\alpha\lambda[1-e^{-\alpha x}]^{\lambda-1}e^{-\alpha x} + bx^{b-1}[1-e^{-\alpha x}]^\lambda\Big]\,dx \quad (2.17)$$

Using series expansion of $e^{-ax^b[1-e^{-\alpha x}]^\lambda}$ as

$$e^{-ax^b[1-e^{-\alpha x}]^\lambda} = \sum_{j=0}^{\infty}\frac{(-1)^j a^j x^{bj}[1-e^{-\alpha x}]^j}{j!} \quad (2.18)$$

We substitute (2.18) in (2.17)

$$E(X^r) = a\int_0^\infty x^r \sum_{j=0}^{\infty} \frac{(-1)^j a^j x^{bj}[1-e^{-\alpha x}]^{\lambda j}}{j!} \left[x^b \alpha\lambda[1-e^{-\alpha x}]^{(\lambda-1)} e^{-\alpha x} + bx^{(b-1)}[1-e^{-\alpha x}]^{\lambda}\right] dx$$

$$= a\sum_{j=0}^{\infty} \frac{(-1)^j a^j}{j!} \left[\alpha\lambda \int_0^\infty x^{r+b(j+1)}[1-e^{-\alpha x}]^{\lambda(j+1)-1} e^{-\alpha x} dx + b\int_0^\infty x^{r+b(j+1)-1}[1-e^{-\alpha x}]^{\lambda(j+1)} dx\right] \qquad (2.19)$$

using binomial series expansion of $[1-e^{-\alpha x}]^{\lambda(j+1)-1}$ and $[1-e^{-\alpha x}]^{\lambda(j+1)}$

$$[1-e^{-\alpha x}]^{\lambda(j+1)-1} = \sum_{n=0}^{\infty} (-1)^n \binom{\lambda(j+1)-1}{n} e^{-\alpha n x} \qquad (2.20)$$

$$[1-e^{-\alpha x}]^{\lambda(j+1)} = \sum_{m=0}^{\infty} (-1)^m \binom{\lambda(j+1)}{m} e^{-\alpha m x} \qquad (2.21)$$

We substitute (2.20) and (2.21) in (2.19) we get

$$E(X^r) = a\alpha\lambda \sum_{j=0}^{\infty}\sum_{n=0}^{\infty} \frac{(-1)^{j+n} a^j}{j!} \binom{\lambda(j+1)-1}{n} \frac{\Gamma(r+b(j+1)+1)}{(\alpha n+\alpha)^{r+b(j+1)+1}} + ab\sum_{j=0}^{\infty}\sum_{m=0}^{\infty} \frac{(-1)^{j+m} a^j}{j!} \binom{\lambda(j+1)}{m} \frac{\Gamma(r+b(j+1))}{(\alpha m)^{r+b(j+1)}}$$

That is the mean can be obtained using (2.15)

When r=1 then $E(X) = \mu$

$$E(X) = \mu$$

$$= a\alpha\lambda \sum_{j=0}^{\infty}\sum_{n=0}^{\infty} \frac{(-1)^{j+n} a^j}{j!} \binom{\lambda(j+1)-1}{n} \frac{\Gamma(b(j+1)+2)}{(\alpha n+\alpha)^{2+b(j+1)}}$$

$$+ ab \sum_{j=0}^{\infty}\sum_{m=0}^{\infty} \frac{(-1)^{j+m} a^j}{j!} \binom{\lambda(j+1)}{m} \frac{\Gamma(b(j+1)+1)}{(\alpha m)^{b(j+1)+1}} \qquad (2.22)$$

Proof b):

$$\mu_r = E[(x-\mu)^r] = \int_0^{\infty} (x-\mu)^r f(X; a, b, \alpha, \lambda)dx \qquad (2.23)$$

Where μ is the mean of $W - EE$ distribution and $f(X; a, b, \alpha, \lambda)$ is the pdf by using the binomial series expansion of $(x-\mu)^r$ yields :

$$\mu_r = E[(x-\mu)^r] = \sum_{s=0}^{r} \binom{r}{s} (-1)^{r-s} \mu^{r-s} \int_0^{\infty} x^s f(X; a, b, \alpha, \lambda)dx \qquad (2.24)$$

$$= \sum_{s=0}^{r} \binom{r}{s} (-1)^{r-s} \mu^{r-s} \acute{\mu}_s \qquad (2.25)$$

Here $\acute{\mu}_s$ represents the sth moment of $W - EE$, by substituting the equation (2.15) by (replacing r by s) into equation (2.25) .

We get the rth moment about the mean of $W - EE$ distribution as follows :

When :

$$\acute{\mu}_s = a\alpha\lambda\sum_{j=0}^{\infty}\sum_{n=0}^{\infty}\frac{(-1)^{j+n}a^j}{j!}\binom{\lambda(j+1)-1}{n}\frac{\Gamma(s+b(j+1)+1)}{(an+\alpha)^{s+b(j+1)+1}}$$

$$+ab\sum_{j=0}^{\infty}\sum_{m=0}^{\infty}\frac{(-1)^{j+m}a^j}{j!}\binom{\lambda(j+1)}{m}\frac{\Gamma(s+b(j+1))}{(am)^{s+b(j+1)}} \qquad (2.26)$$

$$\mu_r = E[(x-\mu)^r]$$

$$= a\alpha\lambda\sum_{j=0}^{\infty}\sum_{n=0}^{\infty}\sum_{s=0}^{r}\binom{r}{s}\frac{(-1)^{j+n+r-s}a^j\mu^{r-s}}{j!}\binom{\lambda(j+1)-1}{n}\frac{\Gamma(s+b(j+1)+1)}{(an+\alpha)^{s+b(j+1)+1}}$$

$$+ab\sum_{j=0}^{\infty}\sum_{m=0}^{\infty}\sum_{s=0}^{r}\binom{r}{s}\frac{(-1)^{j+m+r-s}a^j\mu^{r-s}}{j!}\binom{\lambda(j+1)}{m}\frac{\Gamma(s+b(j+1))}{(am)^{s+b(j+1)}}$$

Therefore ,we can find the variance for a random variable $X\sim W-EE(a,b,\alpha,\lambda)$ by setting r=2 into equation (2.16) as

$$var(x) = E[(x-\mu)^2]$$

$$= a\alpha\lambda\sum_{j=0}^{\infty}\sum_{n=0}^{\infty}\sum_{s=0}^{2}\binom{2}{s}\frac{(-1)^{j+n+2-s}a^j\mu^{2-s}}{j!}\binom{\lambda(j+1)-1}{n}\frac{\Gamma(s+b(j+1)+1)}{(an+\alpha)^{s+b(j+1)+1}}$$

$$+ab\sum_{j=0}^{\infty}\sum_{m=0}^{\infty}\sum_{s=0}^{2}\binom{2}{s}\frac{(-1)^{j+m+2-s}a^j\mu^{2-s}}{j!}\binom{\lambda(j+1)}{m}\frac{\Gamma(s+b(j+1))}{(am)^{s+b(j+1)}} \qquad (2.27)$$

2.5.2 Coefficient of Skewness

The coefficient of skewness **(CS)** of W-EE distribution is given by equation(1.14) .

Let $CS = \frac{A}{B}$ (2.28)

By equation (2.16) , put $r = 3$ then .

$A = E(X-\mu)^3$

$$= a\alpha\lambda\sum_{j=0}^{\infty}\sum_{n=0}^{\infty}\sum_{s=0}^{3}\binom{3}{s}\frac{(-1)^{j+n+3-s}a^j\mu^{3-s}}{j!}\binom{\lambda(j+1)-1}{n}\frac{\Gamma(s+b(j+1)+1)}{(an+\alpha)^{s+b(j+1)+1}}$$

$$+ ab\sum_{j=0}^{\infty}\sum_{m=0}^{\infty}\sum_{s=0}^{3}\binom{3}{s}\frac{(-1)^{j+m+3-s}a^j\mu^{3-s}}{j!}\binom{\lambda(j+1)}{m}\frac{\Gamma(s+b(j+1))}{(am)^{s+b(j+1)}}$$

By equation (2.27)

$B = \left(\sqrt{var(X)}\right)^3 = \left(var(X)\right)^{\frac{3}{2}}$

$$= \left(a\alpha\lambda\sum_{j=0}^{\infty}\sum_{n=0}^{\infty}\sum_{s=0}^{2}\binom{2}{s}\frac{(-1)^{j+n+2-s}a^j\mu^{2-s}}{j!}\binom{\lambda(j+1)-1}{n}\frac{\Gamma(s+b(j+1)+1)}{(an+\alpha)^{s+b(j+1)+1}}\right.$$

$$\left.+ ab\sum_{j=0}^{\infty}\sum_{m=0}^{\infty}\sum_{s=0}^{2}\binom{2}{s}\frac{(-1)^{j+m+2-s}a^j\mu^{2-s}}{j!}\binom{\lambda(j+1)}{m}\frac{\Gamma(s+b(j+1))}{(am)^{s+b(j+1)}}\right)^{\frac{3}{2}}$$

2.5.3 Coefficient of Kurtosis

The coefficient of kurtosis **(CK)** of W-EE distribution is given by equation(1.15) .

Let $CK = \frac{C}{D}$ (2.28)

By equation (2.16) , put $r = 4$ then .

$C = E(X-\mu)^4$

$$= a\alpha\lambda \sum_{j=0}^{\infty}\sum_{n=0}^{\infty}\sum_{s=0}^{4}\binom{4}{s}\frac{(-1)^{j+n+4-s}a^j\mu^{4-s}}{j!}\binom{\lambda(j+1)-1}{n}\frac{\Gamma(s+b(j+1)+1)}{(an+\alpha)^{s+b(j+1)+1}}$$

$$+ ab\sum_{j=0}^{\infty}\sum_{m=0}^{\infty}\sum_{s=0}^{4}\binom{4}{s}\frac{(-1)^{j+m+4-s}a^j\mu^{4-s}}{j!}\binom{\lambda(j+1)}{m}\frac{\Gamma(s+b(j+1))}{(am)^{s+b(j+1)}}$$

By equation (2.27) then

$$D = (var(X))^2$$

$$= \left(a\alpha\lambda \sum_{j=0}^{\infty}\sum_{n=0}^{\infty}\sum_{s=0}^{2}\binom{2}{s}\frac{(-1)^{j+n+2-s}a^j\mu^{2-s}}{j!}\binom{\lambda(j+1)-1}{n}\frac{\Gamma(s+b(j+1)+1)}{(an+\alpha)^{s+b(j+1)+1}} \right.$$

$$\left. + ab\sum_{j=0}^{\infty}\sum_{m=0}^{\infty}\sum_{s=0}^{2}\binom{2}{s}\frac{(-1)^{j+m+2-s}a^j\mu^{2-s}}{j!}\binom{\lambda(j+1)}{m}\frac{\Gamma(s+b(j+1))}{(am)^{s+b(j+1)}} \right)^2$$

2.5.4 Coefficient of Variation

The coefficient of variance (CV) of W-EE distribution is given by equation (1.16).

Let $CV = \frac{E}{F}$ (2.29)

Such that by equation (2.27) $E = \sqrt{var(X)} = (var(X))^{\frac{1}{2}}$

$$E = (var(X))^{\frac{1}{2}}$$

$$= \left[a\alpha\lambda \sum_{j=0}^{\infty}\sum_{n=0}^{\infty}\sum_{s=0}^{2} \binom{2}{s} \frac{(-1)^{j+n+2-s} a^j \mu^{2-s}}{j!} \binom{\lambda(j+1)-1}{n} \frac{\Gamma(s+b(j+1)+1)}{(an+\alpha)^{s+b(j+1)+1}} \right.$$

$$\left. + ab \sum_{j=0}^{\infty}\sum_{m=0}^{\infty}\sum_{s=0}^{2} \binom{2}{s} \frac{(-1)^{j+m+2-s} a^j \mu^{2-s}}{j!} \binom{\lambda(j+1)}{m} \frac{\Gamma(s+b(j+1))}{(am)^{s+b(j+1)}} \right]^{\frac{1}{2}}$$

By equation (2.22) then $F = E(X)$

$$F = E(X) = a\alpha\lambda \sum_{j=0}^{\infty}\sum_{n=0}^{\infty} \frac{(-1)^{j+n} a^j}{j!} \binom{\lambda(j+1)-1}{n} \frac{\Gamma(2+b(j+1))}{(\alpha n+\alpha)^{2+b(j+1)}}$$

$$+ ab \sum_{j=0}^{\infty}\sum_{m=0}^{\infty} \frac{(-1)^{j+m} a^j}{j!} \binom{\lambda(j+1)}{m} \frac{\Gamma(1+b(j+1))}{(\alpha n+\alpha)^{1+b(j+1)}}$$

2.6 Order Statistics

In this section , the pdf of the jth order statistic and the pdf of the smallest and largest order statistics of $W-EE$ distribution are derived .

Let $x_1, x_2, \ldots, x_n$ a radom sample from an $W-EE$ distribution and $x_{1:n} \le x_{2:n} \le \cdots \le x_{n:n}$ denote the order statistics obtained from this sample , then the pdf of $x_{j:n}$ is given by

$$f_{j:n}(X, a, b, \alpha, \lambda) = \frac{n!}{(j-1)!\,(n-j)!} f(X, a, b, \alpha, \lambda)[F(X, a, b, \alpha, \lambda)]^{j-1}[1 - F(X, \alpha, \lambda, a, b)]^{n-j}$$

(2.30)

Where $f(X,a,b,\alpha,\lambda)$ is pdf of $W-EE$ distribution given by equation (2.4) , $F(X,a,b,\alpha,\lambda)$ is cdf of $W-EE$ distribution given by equation (2.3).

$$f_{j:n}(X,a,b,\alpha,\lambda)$$
$$= n\left[ae^{-ax^b[1-e^{-\alpha x}]^\lambda}\left[x^b\alpha\lambda e^{-\alpha x}[1-e^{-\alpha x}]^{\lambda-1}\right.\right.$$
$$\left.\left.+\,bx^{b-1}[1-e^{-\alpha x}]^\lambda\right]\right]\left[1-e^{-ax^b[1-e^{-\alpha x}\,]^\lambda}\right]^{j-1}\left[1\right.$$
$$\left.-\left(1-e^{-ax^b[1-e^{-\alpha x}\,]^\lambda}\right)\right]^{n-j} \quad (2.31)$$

Then the pdf of maximum and minimum are explained as follows.

1. When $j=1$, the pdf of minimum.

$$f_{1:n}(X,a,b,\alpha,\lambda)$$
$$= n\left[ae^{-ax^b[1-e^{-\alpha x}]^\lambda}\left[x^b\alpha\lambda e^{-\alpha x}[1-e^{-\alpha x}]^{\lambda-1}\right.\right.$$
$$\left.\left.+\,bx^{b-1}[1-e^{-\alpha x}]^\lambda\right]\right]\left[e^{-ax^b[1-e^{-\alpha x}\,]^\lambda}\right]^{n-1} \quad (2.32)$$

2. When $j=n$, then pdf of maximum .

$$f_{n:n}(X,a,b,\alpha,\lambda)$$
$$= \frac{n!}{(n-1)!}\left[ae^{-ax^b[1-e^{-\alpha x}]^\lambda}\left[x^b\alpha\lambda e^{-\alpha x}[1-e^{-\alpha x}]^{\lambda-1}\right.\right.$$
$$\left.\left.+\,bx^{b-1}[1-e^{-\alpha x}]^\lambda\right]\right]\left[1-e^{-ax^b[1-e^{-\alpha x}\,]^\lambda}\right]^{n-1} \quad (2.33)$$

2.7 Reni Entropy [12]

Renyi entropy is necessary for quantitative information as can be used as a measure of entanglement ,and it is also importantant in statistics and ecology as an indicator diversity. The renyi entropy of the random variable X with pdf is known as follows [12]

$$I_r(\delta) = \frac{1}{1-\delta}\log(\int_0^\infty f^\delta(\alpha)\,dx\ , wher\ \delta > 0, \delta \neq 1$$

$$= \frac{1}{1-\delta}\log(\int_0^\infty \left[ax^b e^{-ax^b[1-e^{-\alpha x}]^\lambda}[1-e^{-\alpha x}]^\lambda \left[\frac{\alpha\lambda}{e^{\alpha x}-1} + \frac{b}{x}\right]\right]^\delta dx)$$
$$, wher\ \delta > 0, \delta \neq 1 \qquad (2.34)$$

From equation (2. 5) we get the renyi entropy of x given eq. (2.34) by applying the same steps for finding $\acute{\mu}_r$.

2.8 Parameters Estimation of W-EE

In this section , the two considered estimation methods (the maximum likelihood estimation and the moment method) are illustrated to estimate the for parameters of W-EE distribution .

2.8.1 Maximum Likelihood Estimation

The maximum likelihood as a method of estimation for parameters is the most commonly used in the literature. In this study, we discuss the maximum likelihood estimation (MLEs) for the parameters of the W-EE distribution for complete samples. For a random samples .

If $x_1, x_2, x_3, \dots\ , x_n$ denoted random sample from the $\boldsymbol{X \sim W - EE}$ distribution , then the likelihood function is given by

$$L = \prod_{i=1}^{n} f(X_i; \alpha, \lambda, a, b) \qquad (2.35)$$

Form substitution the eq.(2.5) into eq.(2.35)

$$L = \prod_{i=1}^{n} \left[ax^b e^{-ax^b[1-e^{-\alpha x}]^\lambda}[1-e^{-\alpha x}]^\lambda \left[\frac{\alpha\lambda}{e^{\alpha x}-1} + \frac{b}{x}\right]\right] \tag{2.36}$$

$$= (a)^n \prod_{i=1}^{n} x_i{}^b \, e^{-a\sum_{i=1}^{n} x_i{}^b[1-e^{-\alpha x_i}]^\lambda} \prod_{i=1}^{n}[1-e^{-\alpha x_i}]^\lambda \left[\frac{\alpha\lambda}{e^{\alpha x_i}-1} + \frac{b}{x_i}\right] \tag{2.37}$$

Take ln

$$l = n\ln a + \sum_{i=1}^{n} \ln x_i{}^b - a\sum_{i=1}^{n} x_i{}^b\,[1-e^{-\alpha x_i}]^\lambda + \lambda\sum_{i=1}^{n}\ln[1-e^{-\alpha x_i}] + \sum_{i=1}^{n}\ln\left[\frac{\alpha\lambda}{e^{\alpha x_i}-1} + \frac{b}{x_i}\right] \tag{2.38}$$

$$\frac{\partial l}{\partial a} = \frac{n}{a} - \sum_{i=1}^{n} x_i{}^b\,(1-e^{-\alpha x_i})^\lambda = 0 \tag{2.39}$$

$$\frac{\partial l}{\partial b} = \sum_{i=1}^{n} lnx_i - a\sum_{i=1}^{n} x_i{}^b\,(1-e^{-\alpha x})^\lambda \ln x_i + \sum_{i=1}^{n}\frac{(e^{\alpha x_i}-1)}{\alpha\lambda x_i + be^{\alpha x_i} - b} = 0 \tag{2.40}$$

$$\frac{\partial l}{\partial \alpha} = -a\sum_{i=1}^{n} x_i{}^{b+1}\lambda[1-e^{-\alpha x_i}]^{\lambda-1}\,e^{-\alpha x_i} + \sum_{i=1}^{n}\lambda\left(\frac{x_i e^{-x_i}}{1-e^{-\alpha x_i}}\right) + \sum_{i=1}^{n}\frac{(\lambda(e^{\alpha x_i}-1) - \alpha\lambda x_i e^{\alpha x_i})}{\left(\frac{\alpha\lambda}{e^{\alpha x_i}-1} + \frac{b}{x_i}\right).(e^{\alpha x_i}-1)^2} \tag{2.41}$$

$$\frac{\partial l}{\partial \lambda} = -a\sum_{i=1}^{n} x_i^{\,b+1}\,(\ln(1-e^{-\alpha x_i}))(1-e^{-\alpha x_i})^{\lambda} + \sum_{i=1}^{n} \ln[1-e^{-\alpha x_i}]$$

$$+\sum_{i=1}^{n}\frac{\alpha x_i}{\alpha\lambda x_i + be^{\alpha x_i} - b} = 0 \qquad (2.42)$$

We can obtain the MLES of the parameters $\alpha, \lambda, a, and\ b$ by solving the equations (2.43)-(2.46) numerically for $\hat{\alpha}, \hat{\lambda}, \hat{a}, and\ \hat{b}$.

2.8.2 The Method of Moment Estimator

Let $\boldsymbol{x_1, x_2, x_3, \dots\dots, x_n}$be random sample from the $\boldsymbol{X \sim W - EE(\alpha, \lambda, a, b)}$ the method of moment of W-EE distribution is defined by the following equation

$$E(X^r) = \sum_{s=1}^{n}\frac{1}{n}{X_s}^{r} \qquad (2.43)$$

Where $E(X^r)$ is the rth moment about origin given equation (2.15).

For the case $r = 1$, equation (2.43) becomes as follows:

$$E(X^1) = a\alpha\lambda\sum_{j=0}^{\infty}\sum_{n=0}^{\infty}\frac{(-1)^{j+n}a^j}{j!}\binom{\lambda(j+1)-1}{n}\frac{\Gamma(2+b(j+1))}{(\alpha n+\alpha)^{2+b(j+1)}}$$

$$+ab\sum_{j=0}^{\infty}\sum_{m=0}^{\infty}\frac{(-1)^{j+m}a^j}{j!}\binom{\lambda(j+1)}{m}\frac{\Gamma(1+b(j+1))}{(\alpha m)^{1+b(j+1)}} = \sum_{s=1}^{n}\frac{1}{n}{X_s}^{1}$$

$$= \bar{X} \qquad (2.44)$$

For the case $r = 2$, equation (2.43) becomes as follows:

$$E(X^2) = a\alpha\lambda \sum_{j=0}^{\infty}\sum_{n=0}^{\infty} \frac{(-1)^{j+n} a^j}{j!} \binom{\lambda(j+1)-1}{n} \frac{\Gamma(3+b(j+1))}{(\alpha n+\alpha)^{3+b(j+1)}}$$

$$+ ab \sum_{j=0}^{\infty}\sum_{m=0}^{\infty} \frac{(-1)^{j+m} a^j}{j!} \binom{\lambda(j+1)}{m} \frac{\Gamma(2+b(j+1))}{(\alpha m)^{2+b(j+1)}}$$

$$= \sum_{s=1}^{n} \frac{1}{n} {X_s}^2 \tag{2.45}$$

For the case $r = 3$, equation (2.43) becomes as follows:

$$E(X^3) = a\alpha\lambda \sum_{j=0}^{\infty}\sum_{n=0}^{\infty} \frac{(-1)^{j+n} a^j}{j!} \binom{\lambda(j+1)-1}{n} \frac{\Gamma(4+b(j+1))}{(\alpha n+\alpha)^{4+b(j+1)}}$$

$$+ ab \sum_{j=0}^{\infty}\sum_{m=0}^{\infty} \frac{(-1)^{j+m} a^j}{j!} \binom{\lambda(j+1)}{m} \frac{\Gamma(3+b(j+1))}{(\alpha m)^{3+b(j+1)}}$$

$$= \sum_{s=1}^{n} \frac{1}{n} {X_s}^3 \tag{2.46}$$

For the case $r = 4$, equation (2.43) becomes as follows:

$$E(X^4) = a\alpha\lambda \sum_{j=0}^{\infty}\sum_{n=0}^{\infty} \frac{(-1)^{j+n} a^j}{j!} \binom{\lambda(j+1)-1}{n} \frac{\Gamma(5+b(j+1))}{(\alpha n+\alpha)^{5+b(j+1)}}$$

$$+ ab \sum_{j=0}^{\infty}\sum_{m=0}^{\infty} \frac{(-1)^{j+m} a^j}{j!} \binom{\lambda(j+1)}{m} \frac{\Gamma(4+b(j+1))}{(\alpha m)^{4+b(j+1)}}$$

$$= \sum_{s=1}^{n} \frac{1}{n} {X_s}^4 \tag{2.47}$$

We can obtain the estimates for the parameters α, λ, a *and* b by solving the equations $(2.44) and (2.47)$for α, λ, a *and* b by using the numerical methods such as Newton Raphson method .

2.9 Applications

In this part , we explain the real data to show the importance of the $W-EE$ distribution , so that we will compare the $W-EE(X; a, b, \alpha, \lambda)$ distribution with the following distributions .

Exponentiated-Weibull (E-W) distributions [19] with cdf

$$F(X; \alpha, \beta, \theta) = \left[1 - e^{(-\lambda x)^{\theta}}\right]^{\alpha} \quad , x > 0$$

Odd Generalized–Exponential Exponential(OGE-E)distributions [20] with cdf

$$F(X; a, \alpha, \lambda) = \left[1 - e^{-\lambda(e^{ax}-1)}\right]^{\alpha} \quad , x > 0$$

Flexible – Weibull (F-W) distributions [21] with cdf

$$F(X; \alpha, \lambda, \beta, \theta) = \left[1 - e^{-e^{(\beta x^{\lambda} + \theta x^{\alpha})}}\right] \quad , x > 0$$

In the order the $W-EE$ distribution with the above distributions the measures of goodness-of-fit including the Akaike Information criterion (AIC) , Hannan-Quinn Information Criterion (HQIC) , Corrected Akaike Information criterion (CAIC) , and Bayesian Information criterion (BIC) are used [13] where :

$$AIC = -2\hat{\ell} + 2q \quad , \quad BIC = -2\hat{\ell} + q\log(n)$$

$$CAIC = -2\hat{\ell} + \frac{2qn}{n-q-1} \quad , \quad HQIC = -2\hat{\ell} + 2q\,\log(\log(n))$$

Where $\hat{\ell}$ denotes the log-likelihood function evaluated at the maximum likelihood estimates , n is the sample size and q is the number of parameters , In general , the distribution , which gives the smallest values from the criteria , shows the better fit to the data .

2.9.1 First Data Set

We have a real data set corresponding to remission time (in months) of random sample of **128** for sick bladder cancer patients [14] **.**

The first data set are :

[0.08,2.09,3.48,4.87,6.94,8.66,13.11,23.63,0.20,2.23,3.52,4.98,6.97,9.02,13.29,0.40,2.26,3.57,5.06,7.09,9.22,13.80,25.74,0.50,2.46,3.64,5.09,7.26,9.47,14.24,25.82,0.51,2.54,3.70,5.17,7.28,9.74,14.76,26.31,0.81,2.62,3.82,5.32,7.32,10.06,14.77,32.15,2.64,3.88,5.32,7.39,10.34,14.83,34,26,0.90,2.69,4.18,5.34,7.59,10.66,15.96,36.66,1.05,2.69,4.23,5.41,7.62,10.75,16.62,43.01,1.19,2.75,4.26,5.41,7.63,17.12,46.12,1.26,2.83,4.33,5.49,7.66,11.25,17.14,79.05,1.35,2.87,5.62,7.87,11.64,17.36,1.40,3.02,4.34,5.71,7.93,11.79,18.10,1.46,4.40,5.85,8.26,11.98,19.13,1.76,3.25,4.50,6.25,8.37,12.02,2.02,3.31,4.51,6.54,8.53,12.03,20.28,2.02,3.36,6.76,12.07,21.73,2.07,3.36,6.93,8.65,12.63,22.69].

The descriptive statistics for the first data in Table (2.1) , the MLEs of the model parameters estimates and maximum likelihood estimation **($\hat{\ell}$)** for the first data are given in Table (2.2) and the numerical values of the model selection statistics AIC, HQIC, CAIC and BIC with advantage criterion are listed in Table (2.3) .

We can see from table (2.3) that the $W-EE$ model gives the smallest values for the criteria AIC, HQIC, CAIC and BIC so it represents the first data set better than the other selected models , and the criteria AIC is advantage criterion.

Table2.1. descriptive statistics for the first data

Mean	9.3656
Median	6.3950
Variance	110.425
Std. Deviation	10.50833
Minimum	0.08
Maximum	79.05
Range	78.97
Skewness	3.326
Kurtosis	16.154

Table2.2. parameters estimates and Maximum Likelihood estimation ($\hat{\ell}$) for the first data

Model	**Parameters estimates**				$\hat{\ell}$
W-EE(X;a,b,α,λ)	$\hat{a} = 0.111$	$\hat{b} = 1.015$	$\hat{\alpha} = 0.413$	$\hat{\lambda} = 0.436$	-415.2734
E-W(X;a,α,λ)	$\hat{a} = 1.422$	$\hat{\lambda} = 0.192$	$\hat{\theta} = 0.79$	_	-420.655
OGE-E(X;a,α,λ)	$\hat{a} = 0.1$	$\hat{\alpha} = 0.1$	$\hat{\lambda} = 0.1$	_	-7935505
F-W (X;α,λ,β,θ)	$\hat{\alpha} = 0.1$	$\hat{\beta} = 0.1$	$\hat{\lambda} = 0.1$	$\hat{\theta} = 0.539$	-739.4106

Table 2.3. The values of statistics AIC, HQIC, CAIC and BIC with advantage criterion for the first data set

Model	AIC	HQIC	CAIC	BIC	advantage criterion
W-EE(X;a,b,α,λ)	838.5469	849.9550	838.8721	843.1820	838.5469
E-W(X;a,α,λ)	847.2665	860.6747	847.4601	850.7429	847.2665
OGE-E(X;a,α,λ)	1593.1	1606.5	1593.3	1596.6	1593.1
F-W (X;α,λ,β,θ)	1486.8	1498.2	1487.1	1491.5	1486.8

2.9.2 Second Data Set

We have a real data set from Colorado Climate Center, Colorado State University (http://ulysses.atmos.colostate.edu). These data consist of **100** annual maximum precipitation (inches) for one rain gauge in Fort Collins, Colorado, from 1900 through 1999. The data set are : [239, 232, 434, 85, 302, 174, 170, 121, 193, 168, 148, 116, 132, 132, 144, 183, 223, 96, 298, 97, 116, 146, 84, 230, 138, 170, 117, 115, 132, 125, 156, 124, 189, 193, 71, 176, 105, 93, 354, 60, 151, 160, 219, 142, 117, 87, 223, 215, 108, 354, 213, 306, 169, 184, 71, 98, 96, 218, 176, 121, 161, 321, 102, 269, 98, 271, 95, 212, 151, 136, 240, 162, 71, 110, 285, 215, 103, 443, 185, 199, 115, 134, 297, 187, 203, 146, 94, 129, 162, 112, 348, 95, 249, 103, 181, 152, 135, 463, 183, 241].

The descriptive statistics for the first data in Table (2.4) , the MLEs of the model parameters estimates and maximum likelihood estimation $(\hat{\ell})$ for the first data are given in Table (2.5) and the numerical values of the model selection statistics AIC, HQIC, CAIC and BIC with advantage criterion are listed in Table (2.6) .

We can see from table (2.6) that the $W-EE$ model gives the smallest values for the criteria AIC, HQIC, CAIC and BIC so it represents the first data set better than the other selected models , and the criteria AIC is advantage criterion.

Table2.4. descriptive statistics for the second data

Mean	175.6700
Median	158.0000
Variance	6916.728
Std. Deviation	83.16687
Minimum	60.00
Maximum	463.00
Range	403.00
Skewness	1.357
Kurtosis	1.966

Table 2.5. parameters estimates and Maximum Likelihood estimation ($\hat{\ell}$) for the second data

Model	Parameters estimates				$\hat{\ell}$
W-EE(X;a,b,α,λ)	$\hat{a}=0.018$	$\hat{b}=0.777$	$\hat{\alpha}=0.501$	$\hat{\lambda}=1.123$	-638.183
E-W(X;a,α,λ)	$\hat{a}=8.086$	$\hat{\lambda}=0.695$	$\hat{\theta}=0.204$	_	-678.3209
OGE-E(X;a,α,λ)	$\hat{a}=1.016$	$\hat{\alpha}=0.776$	$\hat{\lambda}=0.232$	_	-1858.8
F-W F(X;α,λ,β,θ)	$\hat{\alpha}=0.02$	$\hat{\beta}=0.615$	$\hat{\lambda}=0.162$	$\hat{\theta}=0.02$	-930.4775

Table 2.6. The values of statistics AIC, HQIC , CAIC and BIC with advantage criterion for the Second Data Set

Model	AIC	HQIC	CAIC	BIC	advantage criterion
W-EE(X;a,b,α,λ)	1284.4	1288.6	1284.8	1294.8	1284.4
E-W(X;a,α,λ)	1362.6	1365.8	1362.9	1375.1	1362.6
OGE-E(X;a,α,λ)	3723.7	3726.8	3723.9	3736.1	3723.7
F-W (X;α,λ,β,θ)	1869	1869.4	1869.4	1879.4	1869

From the table (2,1),(2.2),(2,3),(2,4),(2.5) and (2,6) find that our distribution (W-EE) is better than the distributions that have been compared with it and through the different pdf forms we have obtained to distribute us to note that he has great elasticity to be suitable for representing many kinds of different data this flexibility we got after the circular we have done and the table (2.3) and (2.6) it gives the smallest values for criteria AIC , which means that the criteria AIC is advantage criterion relative to the other criteria .

Chapter Three

Modeling for the Lifetime Data with Rayleigh-Exponentiated Exponential Distribution $(R - EE)$

3.1 Introduction

In this chapter a new distribution , inducing a new three- parameters continuous distribution called the Rayleigh-Exponentiated Exponential distribution (**R − EE**) for modeling life time data that created in a new way by using two life distributions .

The Rayleigh-Exponentiated Exponential distribution has been used to model four lifetime data sets and the results show that the proposed generalization performs better than the other known extensions of the Rayleigh distribution considered for the study.

We studied the three parameters Rayleigh - Exponentiated Exponential distribution $R - EE\ (a, b, \lambda)$ distribution will be studied.

3.2 The Cdf and Pdf of $(R - EE)$ Distribution

The cdf of the Rayleigh distribution is :

$$F(x; \lambda) = \left[1 - e^{\frac{-x^2}{2\lambda^2}}\right] \ ,\ x > 0\ , \lambda > 0 \qquad (3.1)$$

Therefore cdf the Exponentiated exponential distribution function is as :

$$F(X; a, b) = [1 - e^{-ax}\ ]^b\ \ ,\ x > 0\ , a\ , b > 0 \qquad (3.2)$$

We multiply $(\boldsymbol{x^2})$ in equation (3.1) by equation (3.2) then we get the cdf $(R - EE)$ distribution as the following equation:

$$F(X; a, b, \lambda) = 1 - e^{\frac{-x^2[1-e^{-ax}]^b}{2\lambda^2}} , x > 0, , a, b, \lambda > 0 \qquad (3.3)$$

Then the pdf of $R - EE$ is given by :

$$f(X;a,b,\lambda) = \left(\frac{1}{2\lambda^2} e^{-\frac{x^2(1-e^{-ax})^b}{2\lambda^2}}\right) (abx^2(1-e^{-ax})^{b-1}e^{-ax}$$
$$+ 2x(1-e^{-ax})^b) \qquad (3.4)$$

$$f(X;a,b,\lambda) = \left[\left(\frac{x[1-e^{-ax}]^b}{\lambda^2} e^{\frac{-x^2[1-e^{-ax}]^b}{2\lambda^2}}\right)\left(\frac{abx}{2[e^{ax}-1]} + 1\right)\right], x > 0 \; , \lambda, a, b > 0 \qquad (3.5)$$

Where b are shape parameters and a, λ is scale parameters.

Now to prove $f(x;a,b,\lambda)$ it is pdf , we must prove that:

1. $f(x;a,b,\lambda) \geq 0$
2. $\int_0^\infty f(X;a,b,\lambda)dx = 1$

That is

1. $f(X;a,b,\lambda) \geq 0$

In (R) distribution $x \in [0,\infty]$and (EE) distribution $x \in [0,\infty)$

then in (R-EE) distribution $x \in [0,\infty]$
When $x \to 0$, then $f(X;a,b,\lambda) = 0$ and as $x \to \infty$, then $f(x;a,b,\lambda) = 0$

$$\therefore f(x;\alpha,\lambda,a,b) \geq 0$$

2. $\int_0^\infty f(X;a,b,\lambda)\,dx = \int_0^\infty \left[\left(\frac{x[1-e^{-ax}]^b}{\lambda^2} e^{\frac{-x^2[1-e^{-ax}]^b}{2\lambda^2}}\right)\left(\frac{abx}{2[e^{ax}-1]} + 1\right)\right] dx$

$$= \left(-e^{\frac{-x^2[1-e^{-ax}]^b}{2\lambda^2}}\right)\Bigg]_0^\infty = 1 \qquad (3.7)$$

The plots of the cdf $F(X; a, b, \lambda)$ are given by the following figures

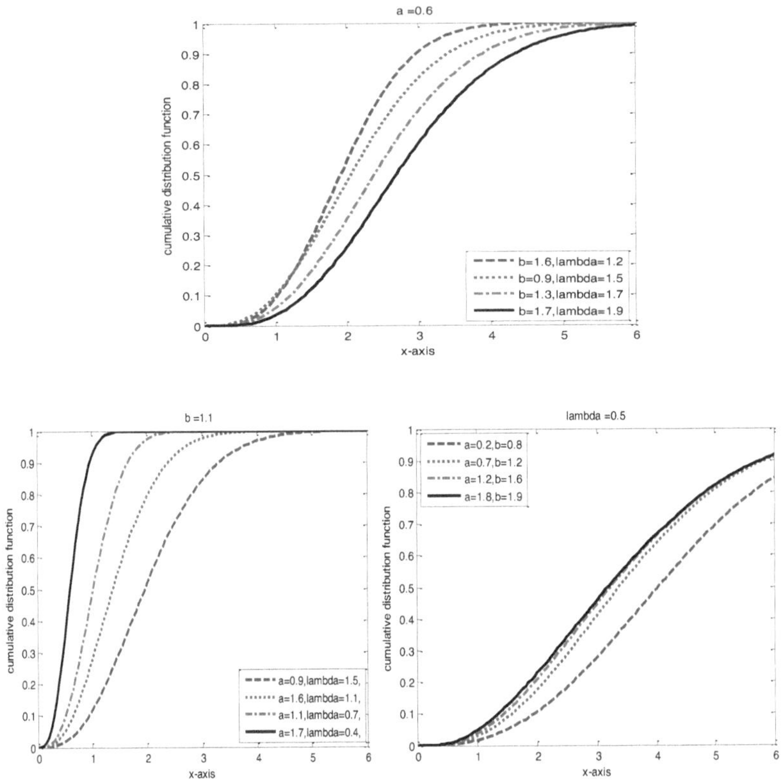

Figure (3.7) : The cdf of $R - EE$ distribution , that the cdf is non-decreasing with increasing x and the parameters a,b,λ

The plots of the probability density function $f(X; a, b, \alpha, \lambda)$ are given by the following figures

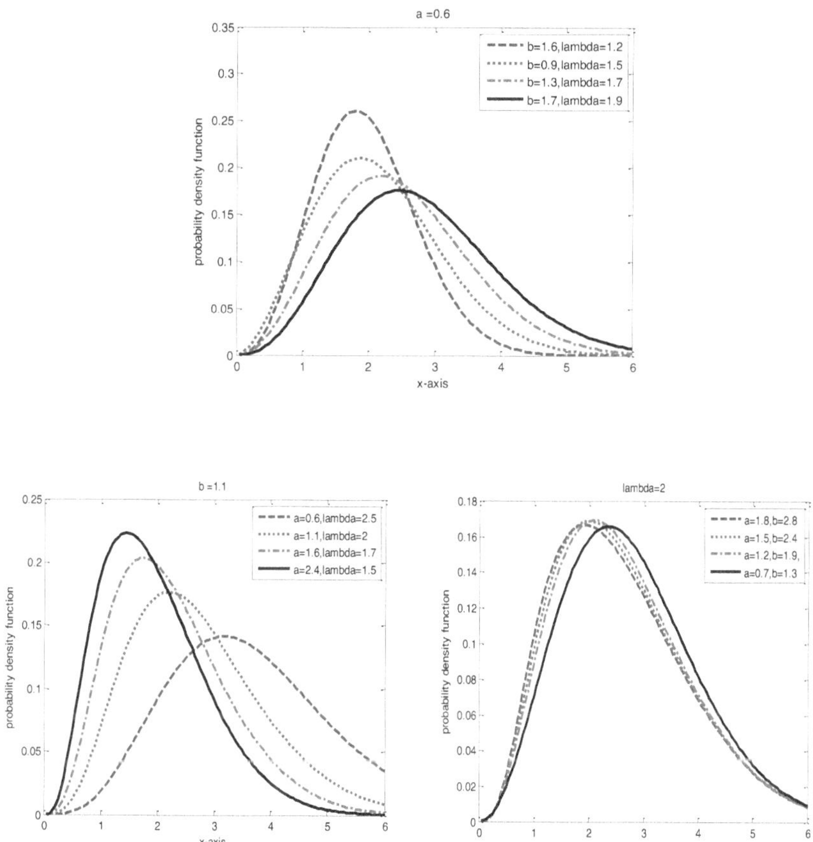

Figure(3.8): The pdf of $R - EE$ distribution , indicate that the R-EE family generates various shapes such as symmetrical ,left-skewed , right skewed ,and reversed –J

3.3 The Limit of The Pdf and Cdf of R-EE

The limit of the pdf is given as follows :

$$\lim_{x\to 0} f(X;a,b,\lambda) = \lim_{x\to 0}\left[\left(\frac{x[1-e^{-ax}]^b}{\lambda^2}\, e^{\frac{-x^2[1-e^{-ax}]^b}{2\lambda^2}}\right)\left(\frac{abx}{2[e^{ax}-1]}+1\right)\right] = 0 \quad (3.8)$$

$$\lim_{x\to \infty} f(X;a,b,\lambda) = \lim_{x\to \infty}\left[\left(\frac{x[1-e^{-ax}]^b}{\lambda^2}\, e^{\frac{-x^2[1-e^{-ax}]^b}{2\lambda^2}}\right)\left(\frac{abx}{2[e^{ax}-1]}+1\right)\right] = 0 \quad (3.9)$$

This is the pdf approaches to 0 as x approach to 0 or ∞ and that is shown Figure(3.8)

The limit of cdf is given as follows :

$$\lim_{x\to 0} F(X;a,b,\lambda) = \lim_{x\to 0}\left[1-e^{\frac{-x^2[1-e^{-ax}]^b}{2\lambda^2}}\right] = 0 \qquad (3.10)$$

$$\lim_{x\to \infty} F(X;a,b,\lambda) = \lim_{x\to \infty}\left[1-e^{\frac{-x^2[1-e^{-ax}]^b}{2\lambda^2}}\right] = 1 \qquad (3.11)$$

This is the cdf approaches to 0 as x approach to 0 and the cdf approaches to 1 as x approach to ∞ and that is shown figure (3.7) .

3.4 Reliability Functions

In this section , we introduce the reliability (survival) function $\bar{F}(X)$, hazard function $h(X)$ and cumulative hazard rate function $H(X)$ and reverse cumulative hazard rate $r(X)$ of $\boldsymbol{X \sim R-EE(a,b,\lambda)}$.

3.4.1 Reliability Function

The reliability (survivor) function of $\boldsymbol{X \sim R-EE(a,b,\lambda)}$ is defined as follows .

$$\bar{F}(X;a,b,\lambda) = 1 - F(X;\alpha,\lambda,a,b) = \left[e^{\frac{-x^2[1-e^{-ax}]^b}{2\lambda^2}}\right] \qquad (3.12)$$

The plots of the reliability function $\bar{F}(x; a, b, \alpha, \lambda)$ are given by the following figures .

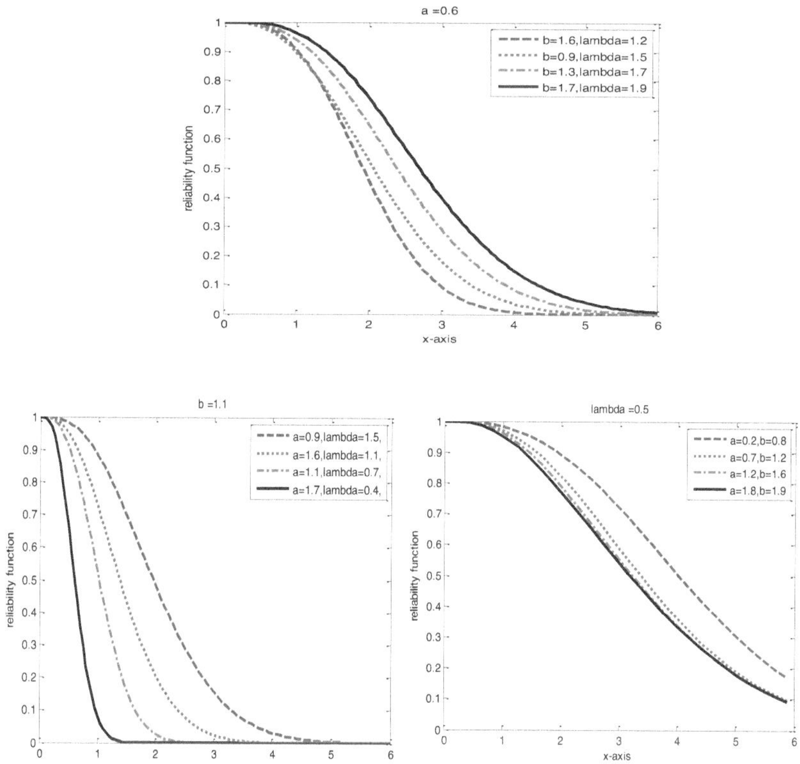

Figure (3.9) : The $\bar{F}(X)$ of $R-EE$ distribution , that the cdf of R-EE is decreasing with decreasing x and the parameters α , λ , a , b.

3.4.2 Hazard Function

The hazard function of $\boldsymbol{X \sim R - EE(a, b, \lambda)}$ is as follows :

$$h(X; a, b, \lambda) = \frac{f(X; a, b, \lambda)}{\bar{F}(X; a, b, \lambda)}$$

$$= \frac{\left[\left(\frac{x[1-e^{-ax}\ ]^{b}}{\lambda^{2}}\ e^{\frac{-x^{2}[1-e^{-ax}]^{b}}{2\lambda^{2}}}\right)\left(\frac{abx}{2[e^{ax}-1]}+1\right)\right]}{\left[e^{\frac{-x^{2}[1-e^{-ax}]^{b}}{2\lambda^{2}}}\right]} \tag{3.13}$$

The plots of the hazard function$h(X;a,b,\lambda)$ are given by the following figures .

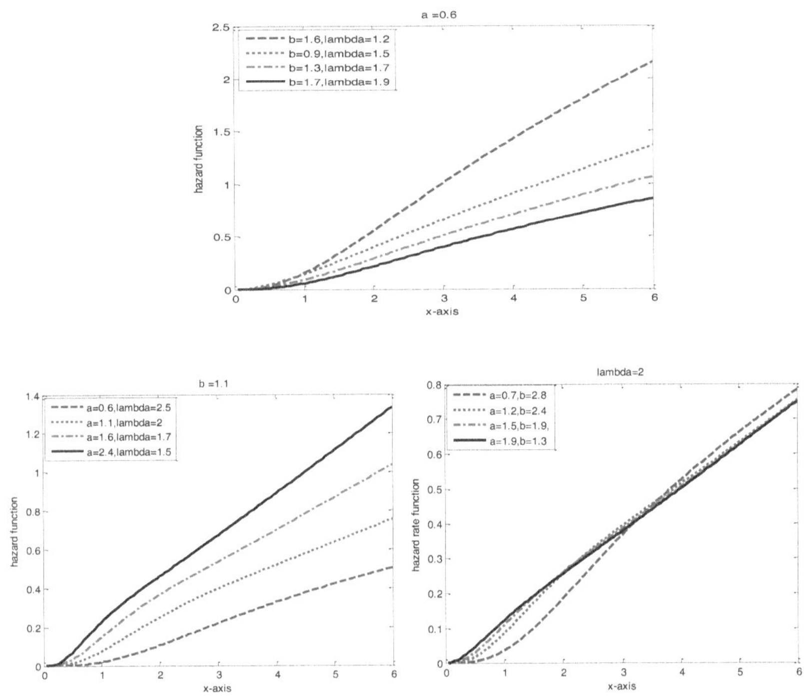

Figure (3.10) : The hazard function of $R-EE$ distribution , indicate that the $h(X)$ of $W-EE$ is non-decreasing function , with creasing x the function creasing .

3.4.3 The Reverse Hazard Function [5]

It is :

$$r(x, \alpha, \lambda, a, b) = \frac{f(x; a, b, \lambda)}{F(x; a, b, \lambda)}$$

$$= \frac{\left[\left(\frac{x[1 - e^{-ax}\ ]^b}{\lambda^2}\ e^{\frac{-x^2[1-e^{-ax}]^b}{2\lambda^2}}\right)\left(\frac{abx}{2[e^{ax} - 1]} + 1\right)\right]}{\left[1 - e^{\frac{-x^2[1-e^{-ax}]^b}{2\lambda^2}}\right]} \qquad (3.14)$$

The plots of the reverse hazard function $r(X; a, b, \lambda)$ are given by the following figures .

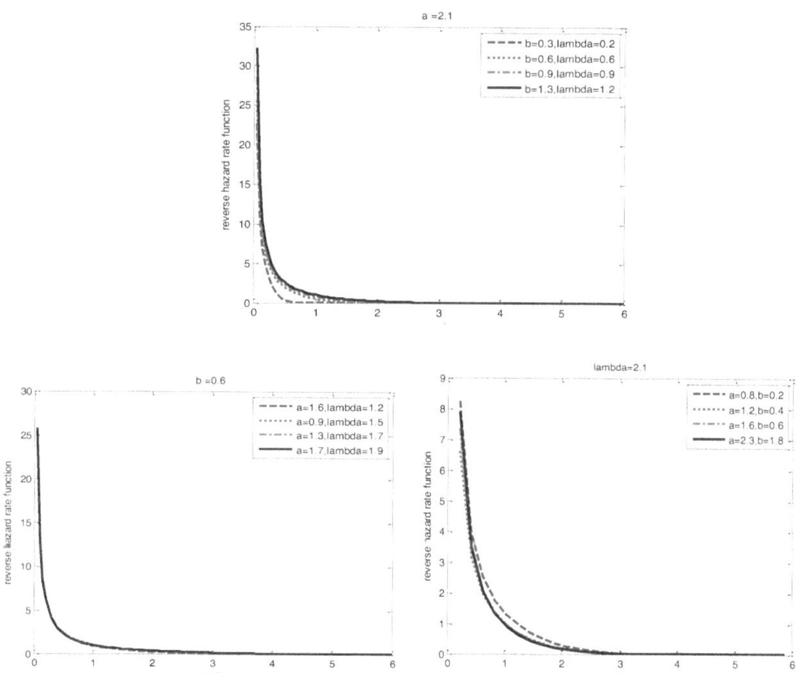

Figure (3.11) : The reverse hazard function indicate that the $r(x)$ of $R - EE$ is decreasing function , when changing x values and parameter

3.4.4 The Cumulative Hazard Function

The chf of $\boldsymbol{X \sim R - EE}(a, b, \lambda)$ is d:

$$H(x; \alpha, \lambda, a, b) = -\ln[1 - F(x, \alpha, \lambda, a, b)] = \frac{x^2[1-e^{-ax}]^b}{2\lambda^2} \tag{3.15}$$

In the following figure, its shape is explained .

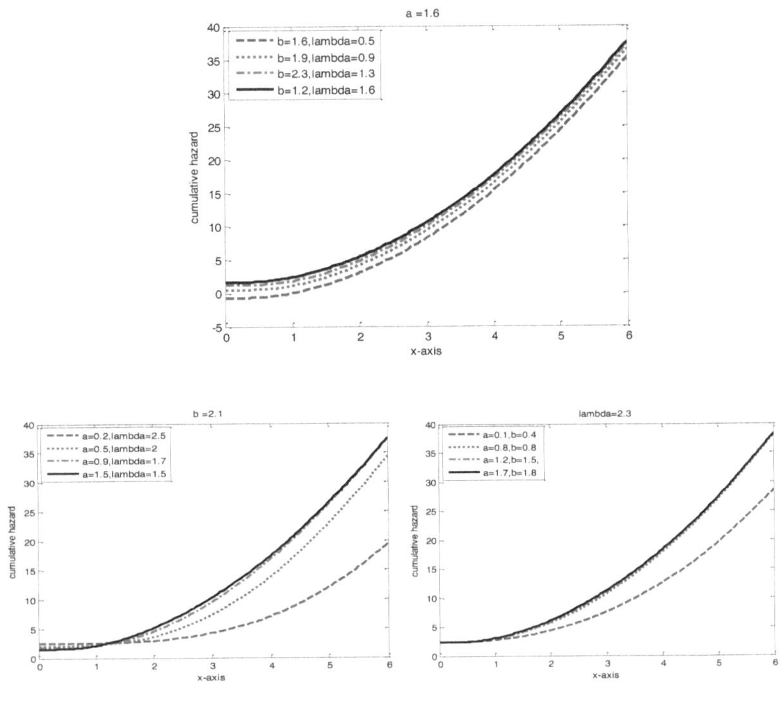

Figure (12) : The cumulative hazard function of R-EE distribution , indicate that the $H(x)$ of $R - EE$ is increasing function when increase x value

3.5 The Moments and Coefficients of Skewness , Kurtosis and Variation

In this section , we introduce the r^{th}moment about the origin , r^{th} moment about the mean , coefficient of skewness , kurtosis and variation for the $R-EE$ distribution .

3.5.1 Theorem

Let $X \sim R-EE(a,b,\lambda)$ the rth moment about the origin , mean respectively are as :

a)

$$\acute{\mu}_r = E(X^r)$$

$$= ab\sum_{i=0}^{\infty}\sum_{m=0}^{\infty}\frac{(-1)^{i+m}}{(2)^{i+1}(\lambda)^{2i+2}i!}\binom{b(i+1)-1}{m}\frac{\Gamma(r+2(i+1)+1)}{(a(m+1))^{r+2(i+1)+1}}$$

$$+2\sum_{i=0}^{\infty}\sum_{n=0}^{\infty}\frac{(-1)^{i+n}}{(2)^{i+1}(\lambda)^{2i+2}i!}\binom{b(i+1)}{n}\frac{\Gamma(r+2i+2)}{(an)^{(r+2i+2)}},r\in Z^{+}$$

(3.16)

b) $\mu_r = E[(X-\mu)^r]$

$$= ab\sum_{i=0}^{\infty}\sum_{m=0}^{\infty}\sum_{s=0}^{r}\binom{r}{s}\frac{(-1)^{i+m+r-s}\mu^{r-s}}{(2)^{i+1}(\lambda)^{2i+2}i!}\binom{b(i+1)-1}{m}\frac{\Gamma(s+2(i+1)+1)}{\left(a(m+1)\right)^{r+2(i+1)+1}}$$

$$+2\sum_{i=0}^{\infty}\sum_{n=0}^{\infty}\sum_{=0}^{r}\binom{r}{s}\frac{(-1)^{i+n+r-s}\mu^{r-s}}{(2)^{i+1}(\lambda)^{2i+2}i!}\binom{b(i+1)}{n}\frac{\Gamma(s+2i+2)}{(an)^{(r+2i+2)}},s \in Z^{+}$$

(3.17)

Proof a):

$$\acute{\mu}_r = E(X^r) = \int_0^{\infty} x^r f(x)\,dx$$

$$\int_0^{\infty}\left[x^r\left(\frac{1}{2\lambda^2}e^{-\frac{x^2(1-e^{-ax})^b}{2\lambda^2}}[(abx^2(1-e^{-ax})^{b-1}e^{-ax}\right.\right.$$

$$\left.\left.+2x(1-e^{-ax})^b]\right)\right]dx \quad (3.18)$$

$$=\int_0^{\infty}\frac{x^r}{2\lambda^2}e^{-\frac{x^2(1-e^{-ax})^b}{2\lambda^2}}[(abx^2(1-e^{-ax})^{b-1}e^{-ax}$$

$$+2x(1-e^{-ax})^b]\,dx \quad (3.19)$$

By using series expansion of $e^{-\frac{x^2(1-e^{-ax})^b}{2\lambda^2}}$ yields .

$$e^{-\frac{x^2(1-e^{-ax})^b}{2\lambda^2}}=\sum_{i=0}^{\infty}\frac{(-1)^i a^j x^{2i}[1-e^{-\alpha x}]^{bi}}{2^i\lambda^{2i}i!} \quad (3.20)$$

We substitute (3.20) in (3.19) we get

$$E(X^r)=\sum_{i=0}^{\infty}\frac{(-1)^i}{2^{i+1}\lambda^{2i+2}i!}\int_0^{\infty}x^{r+2i}[1-e^{-\alpha x}]^{bi}[(abx^2(1-e^{-ax})^{b-1}e^{-ax}+2x(1-e^{-ax})^b]\,dx \quad (3.21)$$

$$=\sum_{i=0}^{\infty}\frac{(-1)^i}{2^{i+1}\lambda^{2i+2}i!}\left[\int_0^{\infty}ab\,x^{r+2(i+1)}[1-e^{-\alpha x}]^{b(i+1)-1}e^{-ax}dx\right.$$

$$\left.+2\int_0^{\infty}x^{r+2i+1}[1-e^{-\alpha x}]^{b(i+1)}\,dx\right] \quad (3.22)$$

using binomial series expansion of $[1-e^{-\alpha x}]^{b(i+1)-1}$ and $[1-e^{-\alpha x}]^{b(i+1)}$

$$[1-e^{-\alpha x}]^{b(i+1)-1}=\sum_{m=0}^{\infty}(-1)^m\binom{b(i+1)-1}{m}e^{-a(m+1)x}$$

(3.23)

$$[1-e^{-\alpha x}]^{b(i+1)} = \sum_{n=0}^{\infty}(-1)^n \binom{b(i+1)}{n} e^{-anx} \qquad (3.34)$$

We substitute (3.24) and (3.23) in (3.22) we get

$$E(X^r) = \sum_{i=0}^{\infty}\frac{(-1)^i}{(2)^{i+1}(\lambda)^{2i+2}\, i!}\left[\int_0^{\infty} ab\, x^{r+2(i+1)} \sum_{m=0}^{\infty}(-1)^m \binom{b(i+1)-1}{m} e^{-a(m+1)x}dx + 2\int_0^{\infty} x^{r+2i+1}\sum_{n=0}^{\infty}(-1)^n\binom{b(i+1)}{n} e^{-anx}\, dx\right]$$

(3.25)

$$E(X^r) = ab\sum_{i=0}^{\infty}\sum_{m=0}^{\infty}\frac{(-1)^{i+m}}{(2)^{i+1}(\lambda)^{2i+2} i!}\binom{b(i+1)-1}{m}\frac{\Gamma(r+2(i+1)+1)}{(a(m+1))^{r+2(i+1)+1}} + 2\sum_{i=0}^{\infty}\sum_{n=0}^{\infty}\frac{(-1)^{i+n}}{(2)^{i+1}(\lambda)^{2i+2} i!}\binom{b(i+1)}{n}\frac{\Gamma(r+2i+2)}{(an)^{(r+2i+2)}}$$

That is the mean can be obtained using (3.16) when r=1

$$E(X) = ab\sum_{i=0}^{\infty}\sum_{m=0}^{\infty}\frac{(-1)^{i+m}}{(2)^{i+1}(\lambda)^{2i+2} i!}\binom{b(i+1)-1}{m}\frac{\Gamma(2+2(i+1))}{(a(m+1))^{2+2(i+1)}} + 2\sum_{i=0}^{\infty}\sum_{n=0}^{\infty}\frac{(-1)^{i+n}}{(2)^{i+1}(\lambda)^{2i+2} i!}\binom{b(i+1)}{n}\frac{\Gamma(3+2i)}{(an)^{(3+2i)}} \qquad (3.26)$$

Proof b): the rth moment about the mean (μ_r) of the random variable x with pdf f(x) is given by :

$$\mu_r = E[(x-\mu)^r] = \int_0^\infty (x-\mu)^r f(X; a, b, \lambda) dx \qquad (3.27)$$

Where μ is the mean of $R - EE$ distribution and $f(X; a, b, \lambda)$ is the pdf by using the binomial series expansion of $(x-\mu)^r$ yields :

$$\mu_r = E[(x-\mu)^r] = \sum_{s=0}^{r} \binom{r}{s} (-1)^{r-s} \mu^{r-s} \int_0^\infty x^s f(X; a, b, \lambda) dx \qquad (3.28)$$

$$= \sum_{s=0}^{r} \binom{r}{s} (-1)^{r-s} \mu^{r-s} \mu_s' \qquad (3.29)$$

Here $\acute{\mu}_s$ represents the sth moment of $R - EE$, by substituting the equation (3.16) by (replacing r by s) into equation (3.29) .

We get the rth moment about the mean of $R - EE$ distribution as follows

$$\acute{\mu}_s = ab \sum_{i=0}^{\infty} \sum_{m=0}^{\infty} \frac{(-1)^{i+m}}{(2)^{i+1} (\lambda)^{2i+2} i!} \binom{b(i+1)-1}{m} \frac{\Gamma(s+2(i+1)+1)}{\left(a(m+1)\right)^{s+2(i+1)+1}}$$
$$+ 2 \sum_{i=0}^{\infty} \sum_{n=0}^{\infty} \frac{(-1)^{i+n}}{(2)^{i+1} (\lambda)^{2i+2} i!} \binom{b(i+1)}{n} \frac{\Gamma(s+2i+2)}{(an)^{(s+2i+2)}},$$

$$s \in Z^+ \qquad (3.30)$$

$$\mu_r = E[(x-\mu)^r]$$

$$= ab \sum_{i=0}^{\infty} \sum_{m=0}^{\infty} \sum_{s=0}^{r} \binom{r}{s} \frac{(-1)^{i+m+r-s} \mu^{r-s}}{(2)^{i+1} (\lambda)^{2i+2} i!} \binom{b(i+1)-1}{m} \frac{\Gamma(s+2(i+1)+1)}{\left(a(m+1)\right)^{s+2(i+1)+1}}$$
$$+ 2 \sum_{i=0}^{\infty} \sum_{n=0}^{\infty} \sum_{s=0}^{r} \binom{r}{s} \frac{(-1)^{i+n+r-s} \mu^{r-s}}{(2)^{i+1} (\lambda)^{2i+2} i!} \binom{b(i+1)}{n} \frac{\Gamma(s+2i+2)}{(an)^{(s+2i+2)}},$$

$$s \in Z^+ \qquad (3.31)$$

Therefore ,we can find the variance for a random variable $X \sim R-EE(a,b,\lambda)$ by setting r=2 into equation (3.17) as

$$var(x) = E[(x-\mu)^r]$$

$$= ab\sum_{i=0}^{\infty}\sum_{m=0}^{\infty}\sum_{s=0}^{2}\binom{2}{s}\frac{(-1)^{i+m+2-s}(\mu)^{2-s}}{(2)^{i+1}(\lambda)^{2i+2}i!}\binom{b(i+1)-1}{m}\frac{\Gamma(s+2(i+1)+1)}{\left(a(m+1)\right)^{S+2(i+1)+1}}$$

$$+2\sum_{i=0}^{\infty}\sum_{n=0}^{\infty}\sum_{s=0}^{2}\binom{2}{s}\frac{(-1)^{i+n+2-s}\,\mu^{r-s}}{(2)^{i+1}(\lambda)^{2i+2}i!}\binom{b(i+1)}{n}\frac{\Gamma(s+2i+2)}{(an)^{(S+2i+2)}} \qquad (3.32)$$

3.5.2 Coefficient of Skewness

The coefficient of skewness (**CS**) of R-EE distribution is given by equation (1.14)

Let

$$CS = \frac{A}{B} \qquad (3.33)$$

By equation (3.17) , put $r = 3$ then

$$A = E(X-\mu)^3$$

$$= \left(ab\sum_{i=0}^{\infty}\sum_{m=0}^{\infty}\sum_{s=0}^{3}\binom{3}{s}\frac{(-1)^{i+m+3-s}\mu^{3-s}}{(2)^{i+1}(\lambda)^{2i+2}i!}\binom{b(i+1)-1}{m}\frac{\Gamma(s+2(i+1)+1)}{(a(m+1))^{s+2(i+1)+1}}\right.$$

$$\left.+2\sum_{i=0}^{\infty}\sum_{n=0}^{\infty}\sum_{s=0}^{3}\binom{3}{s}\frac{(-1)^{i+n+3-s}\,\mu^{3-s}}{(2)^{i+1}(\lambda)^{2i+2}i!}\binom{b(i+1)}{n}\frac{\Gamma(s+2i+2)}{(an)^{(s+2i+2)}}\right)$$

By equation (3.32) then

$$B=\left(\sqrt{var(X)}\right)^3=(var(X))^{\frac{3}{2}}$$

$$=\left(ab\sum_{i=0}^{\infty}\sum_{m=0}^{\infty}\sum_{s=0}^{2}\binom{2}{s}\frac{(-1)^{i+m+2-s}\mu^{2-s}}{(2)^{i+1}(\lambda)^{2i+2}i!}\binom{b(i+1)-1}{m}\frac{\Gamma(s+2(i+1)+1)}{(a(m+1))^{S+2(i+1)+1}}\right.$$

$$\left.+2\sum_{i=0}^{\infty}\sum_{n=0}^{\infty}\sum_{s=0}^{2}\binom{2}{s}\frac{(-1)^{i+n+2-s}\ \mu^{r-s}}{(2)^{i+1}(\lambda)^{2i+2}i!}\binom{b(i+1)}{n}\frac{\Gamma(s+2i+2)}{(an)^{(S+2i+2)}}\right)^{\frac{3}{2}}$$

3.5.3 Coefficient of Kurtosis

The coefficient of kurtosis (**CK**) of R-EE distribution is given by equation (1.15)

Let

$$CK=\frac{C}{D} \tag{3.35}$$

By equation (3.17) , put $r = 4$ then .

$$C=E(X-\mu)^4$$

$$=\left(ab\sum_{i=0}^{\infty}\sum_{m=0}^{\infty}\sum_{s=0}^{4}\binom{4}{s}\frac{(-1)^{i+m+4-s}\mu^{4-s}}{(2)^{i+1}(\lambda)^{2i+2}i!}\binom{b(i+1)-1}{m}\frac{\Gamma(s+2(i+1)+1)}{(a(m+1))^{s+2(i+1)+1}}\right.$$

$$\left.+2\sum_{i=0}^{\infty}\sum_{n=0}^{\infty}\sum_{s=0}^{4}\binom{4}{s}\frac{(-1)^{i+n+4-s}\ \mu^{4-s}}{(2)^{i+1}(\lambda)^{2i+2}i!}\binom{b(i+1)}{n}\frac{\Gamma(s+2i+2)}{(an)^{(s+2i+2)}}\right)$$

By equation (3.32)

$$D=(var(X))^2$$

$$= \left(ab \sum_{i=0}^{\infty} \sum_{m=0}^{\infty} \sum_{s=0}^{2} \binom{2}{s} \frac{(-1)^{i+m+2-s} \mu^{2-s}}{(2)^{i+1} (\lambda)^{2i+2} i!} \binom{b(i+1)-1}{m} \frac{\Gamma(s+2(i+1)+1)}{(a(m+1))^{S+2(i+1)+1}} \right.$$

$$\left. + 2 \sum_{i=0}^{\infty} \sum_{n=0}^{\infty} \sum_{s=0}^{2} \binom{2}{s} \frac{(-1)^{i+n+2-s} \mu^{r-s}}{(2)^{i+1} (\lambda)^{2i+2} i!} \binom{b(i+1)}{n} \frac{\Gamma(s+2i+2)}{(an)^{(S+2i+2)}} \right)^2$$

3.5.4 Coefficient of Variation

The coefficient of variance (CV) of R-EE distribution is given by equation (1.16).

Let

$$CV = \frac{E}{F} \tag{3.37}$$

By equation (3.32) then $E = \sqrt{var(X)} = (var(X))^{\frac{1}{2}}$

$$= \left(ab \sum_{i=0}^{\infty} \sum_{m=0}^{\infty} \sum_{s=0}^{2} \binom{2}{s} \frac{(-1)^{i+m+2-s} (\mu)^{2-s}}{(2)^{i+1} (\lambda)^{2i+2} i!} \binom{b(i+1)-1}{m} \frac{\Gamma(s+2(i+1)+1)}{(a(m+1))^{S+2(i+1)+1}} \right.$$

$$\left. + 2 \sum_{i=0}^{\infty} \sum_{n=0}^{\infty} \sum_{s=0}^{2} \binom{2}{s} \frac{(-1)^{i+n+2-s} \mu^{r-s}}{(2)^{i+1} (\lambda)^{2i+2} i!} \binom{b(i+1)}{n} \frac{\Gamma(s+2i+2)}{(an)^{(S+2i+2)}} \right)^{\frac{1}{2}}$$

By equation (3.26) then $F = E(X)$

$$= \left(ab \sum_{i=0}^{\infty} \sum_{m=0}^{\infty} \frac{(-1)^{i+m}}{(2)^{i+1} (\lambda)^{2i+2} i!} \binom{b(i+1)-1}{m} \frac{\Gamma(2+2(i+1))}{(a(m+1))^{2+2(i+1)}} \right.$$

$$\left. + 2 \sum_{i=0}^{\infty} \sum_{n=0}^{\infty} \frac{(-1)^{i+n}}{(2)^{i+1} (\lambda)^{2i+2} i!} \binom{b(i+1)}{n} \frac{\Gamma(3+2i)}{(an)^{(3+2i)}} \right)$$

3.6 Order Statistics

Let $x_1, x_2, \dots, x_n$ a radom sample from an $R-EE$ distribution and $x_{1:n} \leq x_{2:n} \leq \cdots \leq x_{n:n}$ denote their order statistics , then the pdf of $x_{j:n}$ defined as:

$$f_{j:n}(x,a,b,\lambda) = \frac{n!}{(j-1)!(n-j)!} f(x,a,b,\lambda)[F(x,a,b,\lambda)]^{j-1}[1-F(x,a,b,\lambda)]^{n-j} \tag{3.38}$$

We know that $f(X,,a,b,\lambda)$ is pdf of $R-EE$ distribution given by equation (3.4) , $F(X,a,b,\lambda)$ is cdf of $R-EE$ distribution given by equation (3.3).

$$= \frac{n!}{(j-1)!\,(n-j)!}\left[\frac{1}{2\lambda^2} e^{-\frac{x^2(1-e^{-ax})^b}{2\lambda^2}}(abx^2(1-e^{-ax})^{b-1}e^{-ax} + 2x(1-e^{-ax})^b)\right]\left[1-e^{\frac{-x^2[1-e^{-ax}]^b}{2\lambda^2}}\right]^{j-1}\left[e^{\frac{-x^2[1-e^{-ax}]^b}{2\lambda^2}}\right]^{n-j} \tag{3.39}$$

Then:

1. When $j = 1$, the pdf of minimum.

$$f_{1:n}(X,\alpha,\lambda,a,b) = n\left[\frac{1}{2\lambda^2} e^{-\frac{x^2(1-e^{-ax})^b}{2\lambda^2}}(abx^2(1-e^{-ax})^{b-1}e^{-ax} + 2x(1-e^{-ax})^b)\right]\left[e^{\frac{-x^2[1-e^{-ax}]^b}{2\lambda^2}}\right]^{n-1} \tag{3.40}$$

2. When $j = n$, then pdf of maximum .

$$f_{n:n}(X,\alpha,\lambda,a,b)$$
$$= n\left[\frac{1}{2\lambda^2}e^{-\frac{x^2(1-e^{-ax})^b}{2\lambda^2}}(abx^2(1-e^{-ax})^{b-1}e^{-ax} + 2x(1-e^{-ax})^b)\right]\left[1-e^{\frac{-x^2[1-e^{-ax}]^b}{2\lambda^2}}\right]^{n-1}$$

(3.41)

3.7 Reni entropy

Renyi entropy is necessary for quantitative information as can be used as a measure of entanglement ,and it is also importantant in statistics and ecology as an indicator diversity. The renyi entropy is known as follows :

$$I_r(\delta) = \frac{1}{1-\delta}\log(\int_0^\infty f^\delta(\alpha)\,dx \,, wher\ \delta > 0, \delta \neq 1 \quad (3.42)$$

$$= \frac{1}{1-\delta}\log(\int_0^\infty \left[\left(\frac{x[1-e^{-ax}\ ]^b}{\lambda^2}e^{\frac{-x^2[1-e^{-ax}]^b}{2\lambda^2}}\right)\left(\frac{abx}{2[e^{ax}-1]}+1\right)\right]^\delta dx) \,, wher\ \delta > 0, \delta \neq 1$$

(3.43)

From equation (3.43) we get the renyi entropy of x by applying the same steps for finding $\acute{\mu}_r$.

3.8 Parameters Estimation of R-EE

In this section , the two considered estimation methods (the maximum likelihood estimation and the moment method) are illustrated to estimate the for parameters of R-EE distribution .

3.8.1 Maximum Likelihood Estimation

If $x_1, x_2, x_3, \dots \dots, x_n$ denoted random sample from the $\boldsymbol{X \sim R - EE}$ distribution , then the likelihood function is given by:

$$L = \prod_{i=1}^{n} f(X_i, a, b, \lambda) \tag{3.44}$$

Form substitution the equation (3.5) into equation (3.44)

$$L = \prod_{i=1}^{n} \left[\left(\frac{x[1 - e^{-ax}\ ]^b}{\lambda^2} e^{\frac{-x^2[1-e^{-ax}]^b}{2\lambda^2}} \right) \left(\frac{abx}{2[e^{ax} - 1]} + 1 \right) \right] \tag{3.45}$$

$$= \left(\frac{\prod_{i=1}^{n} x_i \prod_{i=1}^{n} [1 - e^{-ax_i}\ ]^b}{\lambda^{2n}} e^{\sum_{i=1}^{n} \frac{-x_i^2[1-e^{-ax_i}]^b}{2\lambda^2}} \right) \cdot \left(\prod_{i=1}^{n} \left[\frac{a\, bx_i e^{ax_i}}{2[e^{ax_i} - 1]} + 1 \right] \right) \tag{3.46}$$

Take ln

$$l = \sum_{i=1}^{n} \ln(x_i) + b \sum_{i=1}^{n} \ln[1 - e^{-ax_i}\ ] - 2n \ln \lambda - \frac{1}{2\lambda^2} \sum_{i=1}^{n} [x_i^2 (1 - e^{-ax_i})^b] + \sum_{i=1}^{n} \ln \left[\frac{a\, bx_i}{[2e^{ax_i} - 2\,]} + 1 \right] \tag{3.47}$$

$$\frac{\partial l}{\partial a} = b \sum_{i=1}^{n} \frac{x_i e^{-ax_i}}{1 - e^{-ax_i}} - \frac{1}{2\lambda^2} \sum_{i=1}^{n} bx_i^{\,3} e^{-ax_i} (1 - e^{-ax_i})^{b-1} + \sum_{i=1}^{n} \frac{bx_i e^{ax_i} - bx_i - abx_i^{\,2} e^{ax_i}}{abx_i e^{ax_i} - abx_i + 2e^{2ax_i} - 4e^{ax_i} + 2} = 0 \tag{3.48}$$

$$\frac{\partial l}{\partial b}=\sum_{i=1}^{n} ln(1-e^{-ax_i})-\sum_{i=1}^{n}\frac{abx_i^{\,2}(1-e^{ax_i})^{b-1}}{2\lambda^2}+\sum_{i=1}^{n}\frac{ax_i}{2e^{ax_i}-2}=0 \quad (3.49)$$

$$\frac{\partial l}{\partial \lambda}=\sum_{i=1}^{n}\frac{x_i^{\,2}(1-e^{-ax_i})^b}{\lambda^3}-\frac{2n}{\lambda}=0 \quad (3.50)$$

We can obtain the estimators of the parameters $\alpha, \lambda, a, and\ b$ numerically solving the equations (48)-(50) for $\hat{\alpha}, \hat{\lambda}, \hat{a}, and\ \hat{b}$.

3.8.2 The Method Of Moment Estimator

Let $\boldsymbol{x_1, x_2, x_3, \dots \dots, x_n}$be random sample from the $\boldsymbol{X \sim R - EE(a, b, \lambda)}$ the method of moment of R-EE distribution is defined by the following equation

$$E(X^r)=\sum_{s=1}^{n}\frac{1}{n}X_s^{\,r} \quad (3.51)$$

Where $E(X^r)$ is the rth moment about origin given equation (3.24).

For the case $r = 1$, equation (3.51) becomes as follows

$$E(X^1)=ab\sum_{i-0}^{\infty}\sum_{m-0}^{\infty}\frac{(-1)^{i+m}}{(2)^{i+1}(\lambda)^{2i+2}i!}\binom{b(i+1)-1}{m}\frac{\Gamma(2+2(i+1))}{(a(m+1))^{2+2(i+1)}}$$
$$+2\sum_{i=0}^{\infty}\sum_{n=0}^{\infty}\frac{(-1)^{i+n}}{(2)^{i+1}(\lambda)^{2i+2}i!}\binom{b(i+1)}{n}\frac{\Gamma(3+2i)}{(an)^{(3+2i)}}$$
$$=\sum_{s=1}^{n}\frac{1}{n}X_s^{\,1}=\bar{X} \quad (3.52)$$

For the case $r = 2$, equation (3.51) becomes as follows:

$$E(X^2) = ab\sum_{i=0}^{\infty}\sum_{m=0}^{\infty}\frac{(-1)^{i+m}}{(2)^{i+1}(\lambda)^{2i+2}i!}\binom{b(i+1)-1}{m}\frac{\Gamma(3+2(i+1))}{(a(m+1))^{3+2(i+1)}}$$

$$+2\sum_{i=0}^{\infty}\sum_{n=0}^{\infty}\frac{(-1)^{i+n}}{(2)^{i+1}(\lambda)^{2i+2}i!}\binom{b(i+1)}{n}\frac{\Gamma(4+2i)}{(an)^{(4+2i)}}$$

$$=\sum_{s=1}^{n}\frac{1}{n}X_s^{\,2} \qquad (3.53)$$

For the case $r = 3$, equation (3.51) becomes as follows:

$$E(X^3) = ab\sum_{i=0}^{\infty}\sum_{m=0}^{\infty}\frac{(-1)^{i+m}}{(2)^{i+1}(\lambda)^{2i+2}i!}\binom{b(i+1)-1}{m}\frac{\Gamma(4+2(i+1))}{(a(m+1))^{4+2(i+1)}}$$

$$+2\sum_{i=0}^{\infty}\sum_{n=0}^{\infty}\frac{(-1)^{i+n}}{(2)^{i+1}(\lambda)^{2i+2}i!}\binom{b(i+1)}{n}\frac{\Gamma(5+2i)}{(an)^{(5+2i)}}$$

$$=\sum_{s=1}^{n}\frac{1}{n}X_s^{\,3} \qquad (3.54)$$

Thus the estimators for the parameters α, λ, a *and* b using the numerical methods such as Newton Raphson method to solve the equations (3.52)-(3.54) .

3.9. Applications

To show the importance of the $R - EE$ distribution we apply it , and compare the $R - EE(X; a, b, \lambda)$ distribution with the following distributions

Weibull – Rayleigh ($W - R$) distribution [21] with cdf

$$F(X; \alpha, \beta, \theta) = \left[1 - e^{-\alpha\left(e^{\frac{\theta}{2}x^2}-1\right)}\right] \quad , x > 0 .$$

Rayleigh – Rayleigh $(W - R)$ distribution [2] with cdf

$$F(X;\theta,\lambda) = \left[e^{\frac{-\theta}{2}\left(-\ln\left(1-e^{\frac{-\lambda x^2}{2}}\right)\right)^2}\right] ,x > 0 .$$

Extended Odd Weibull Rayleigh (EOWR) distribution [1] with cdf

$$F(X;\alpha,\beta,\delta) = \left[1-\left\{1+\beta\left(e^{\delta x^2}-1\right)^{\alpha}\right\}^{\frac{-1}{\beta}}\right] ,x > 0 .$$

In the order the $R - EE$ distribution with the above distributions the measures of goodness-of-fit use the Akaike information criterion (AIC) , Hannan-Quinn information criterion (HQIC) , corrected Akaike information criterion (CAIC) , and Bayesian Information criterion (BIC) are used [7] where :

$$AIC = -2\hat{\ell} + 2q \quad , \quad BIC = -2\hat{\ell} + q\log(n)$$

$$CAIC = -2\hat{\ell} + \frac{2qn}{n-q-1} \quad , \quad HQIC = -2\hat{\ell} + 2q\ \log(\log(n))$$

Where $\hat{\ell}$ denotes the log-likelihood function estimator, n is the sample size and q is the number of parameters . In general , the distribution R-EE gives the smallest values from the criteria to be the better than the other

Data Set

The data set is the failure times of 84 Aircraft Windshield . times are reported in the book "Weibull Models" by Murthy et al. (2004, page 297). This data set is previously studied by Bassiouny et al (2015) .[15] are data set [0.040, 1.866, 2.385, 3.443, 0.301, 1.876, 2.481, 3.467, 0.309, 1.899, 2.610, 3.478, 0.557, 1.911, 2.625, 3.578, 0.943, 1.912, 2.632, 3.595, 1.070, 1.914, 2.646, 3.699,

1.124, 1.981, 2.661, 3.779,1.248, 2.010, 2.688, 3.924, 1.281, 2.038, 2.82,3, 4.035, 1.281, 2.085, 2.890, 4.121, 1.303, 2.089, 2.902, 4.167, 1.432, 2.097, 2.934, 4.240, 1.480, 2.135, 2.962, 4.255, 1.505, 2.154, 2.964, 4.278, 1.506, 2.190, 3.000, 4.305, 1.568, 2.194, 3.103, 4.376, 1.615, 2.223, 3.114, 4.449, 1.619, 2.224, 3.117, 4.485, 1.652, 2.229, 3.166, 4.570, 1.652, 2.300, 3.344, 4.602, 1.757, 2.324, 3.376, 4.663] .

The descriptive statistics for the first data in Table (3.1) , the MLEs of the model parameters estimates and maximum likelihood estimation $(\hat{\ell})$ for the data are given in Table (3.2) and the numerical values of the model selection statistics AIC, HQIC, CAIC and BIC with advantage criterion are listed in Table (3.3) .

We can see from table (3.3) that the $W - EE$ model gives the smallest values for the criteria AIC, HQIC, CAIC and BIC so it represents the data set better than the other selected models , and the criteria BIC is advantage criterion.

Table3.1. descriptive statistics for the data

Mean	**2.5238**
Median	**2.3120**
Variance	**1.329**
Std. Deviation	**1.15264**
Minimum	**0.04**
Maximum	**4.66**
Range	**4.66**
Skewness	**0.047**
Kurtosis	**0.584**

Table 3.2. parameters estimates and Maximum Likelihood estimation ($\hat{\ell}$) for the data

Model	parameters estimates			$\hat{\ell}$
R-EE(a,b,λ)	$\hat{a} = 1.809$	$\hat{b} = 0.103$	$\hat{\lambda} = 0.583$	-57.948
W-R(α,β,θ)	$\hat{\alpha} = 0.235$	$\hat{\beta} = 0.168$	$\hat{\theta} = 1.663$	-86.173
R-R(θ,λ)	$\hat{\theta} = 0.474$	$\hat{\lambda} = 0.1$	–	-148.152
EOWR (α,σ,δ)	$\hat{\alpha} = 0.117$	$\hat{\sigma} = 0.752$	0.102	-68.235

Table 3.3. The values of statistics AIC , HQIC , CAIC and BIC with advantage criterion for the Data Set

Model	AIC	HQIC	CAIC	BIC	best criterion
R-EE(a,b,λ)	110.0935	107.1620	109.7935	98.3703	98.3703
W-R(α,β,θ)	166.2973	163.3657	165.9973	154.5740	154.5740
R-R(θ,λ)	292.2124	290.2580	292.0642	278.4891	278.4891
EOWR (α,σ,δ)	130.2871	127.3556	129.9871	118.5638	118.5638

From the table (3.1),(3.2) and (3.3) find that our distribution (R-EE) is better than the distributions that have been compared with it and through the different pdf forms we have obtained to distribute us to note that he has great elasticity to be suitable for representing many kinds of different data this flexibility we got after the circular we have done and table (3.3) it gives the smallest values for criteria BIC , which means that the criteria BIC is best criterion relative to the other criteria.

Chapter Four

Conclusions and Future Work

4.1. Conclusions

1) In this works, we have defined new two four-parameters continuous distributions called the Weibull-Exponentiated Exponential distribution (W-EE) and Rayleigh-Exponentiated Exponential (R-EE) distribution for modeling life time data by multiplying the variable in cdf of the W distribution with the cdf of the EE distribution to form the cdf of (W-EE) distribution and by using the same way for R-EE distribution .

2) There are explicit expressions for moments, mean, variance, the coefficient of skewness, kurtosis and variation, the moment generating functions, order statistics, and Rényi entropy .

3) The$(W - EE)$ and the $(R - EE)$ distributions represent the data sets better than the other selected distributions since they have the smallest values of criteria like the AIC, CAIC, HQIC and BIC through two applications to real data sets when we compared our proposed distributions with some other distributions by using.

4) In the $(W - EE)$ distribution they have the smallest values for criteria AIC , which means that the criteria AIC is advantage criterion relative to the other criteria and In the $(R - EE)$ distribution they have the smallest values for criteria BIC , which means that the criteria BIC is advantage criterion relative to the other criteria .

4.2. Future Work

1) Finding the asymptotic intervals for the parameters of $(\mathbf{W}-\mathbf{EE})$ and $(\mathbf{R}-\mathbf{EE})$ distribution .

2) Using other estimation methods to estimate the five parameters of the$(\mathbf{W}-\mathbf{EE})$ and $(\mathbf{R}-\mathbf{EE})$ distribution

References :

[1] Almongy, Hisham M., et al. "A new extended rayleigh distribution with applications of COVID-19 data." Results in Physics 23 (2021): 104012.

[2] Al-Noor, N. H., and N. K. Assi. "Rayleigh-Rayleigh Distribution: Properties and Applications." Journal of Physics: Conference Series. Vol. 1591. No. 1. IOP Publishing, (2020).

[3] Al-Wakeel, A. A. S. Notes on Weibull distribution, Journal of Economic and administrative sciences, Baghdad University 18(67) (2012)(260-272).

[4] Al-Zahrani, B. and Sagor, H. The Poisson-Lomax distribution, 37(1)(2014) (223-243).

[5] Block, H. W. Savits ,T. H. and Singh, H., The Reversed Hazard Rate Function , (2009), Cambridge University press.

[6] Bourguignon, M., Silva, R.B. and Cordeiro, G.M. The Weibull–G family of probability distributions, Journal of Data Science 12, 53–68, 2014 .

[7] Chen,J. M., Hawkes, A. G., Scalas, E. and Trinh,M. performance of information Criteria used for model selection of Hawkes process models of financial data arXiv: 1702.06055v2 [q. fin. ST](2017).

[8] Curto, J. D. and Pinto, J. C., The coefficient of variation asymptotic distribution in the case of non-iid random variables, Journal of Applied Statistics36(1)(2009)(2009)(21-32)

[9] Elgarhy., M., Haq, M.A., Ozel, G., Nasir, M.A., “A new exponentiated extended family of distributions with applications”, Gazi University Journal of Science, 30: 101-115, (2017).

[10] Epstein B (1958) The exponential distribution and its role in Life - testing. Ind Qual Control 15: 2-7.

[11] F. Merovci, Transmuted Rayleigh distribution. Austrian Journal of Statistics, 42(1), 21-31, (2013).

[12] Gupta, R. D and Kundu, D., Exponentiated Exponential Family, An alternative to Gamma and Weibull, Biometrical Journal 43, 2001, 117-130.

[13] Gupta, R. D. and Kundu, D., 1999," Generalized Exponential Distributions ", Australian and New Zealand Journal of Statistics , Vol. 41, No.2, pp. 173-188.

[14] Gupta, R. D. and Kundu, D., 2001, "Exponentiated Exponential Family; An Alternative to Gamma and Weibull ", Biometrical Journal, vol. 33, no. 1, pp.117-130.

[15] H. El-Bassiouny, N. F. Abdo and H. S. Shahen, Exponential Lomax Distribution , International Journal of Computer Applications, 121, 2015,.0975 – 8887.

[16] Hassan, A.S., Elgarhy, M., “A new family of exponentiated Weibull-generated distributions”, International Journal of Mathematics and its Applications, 4: 135-148, (2016).

[17] IVÁDY, P., A NOTE ON A GAMMA FUNCTION INEQUALITY, Journal of Mathematical Inequalities, 3(2) (2009) (227-236).

[18] Konnully, A. O., Relation Between the Beta and the Gamma Functions, Mathematics Magazine, 41(1) (2018) (37-39).

[19] Krishnamoorthy, K. (2006), Handbook of Statistical Distribution with Applications , Taylor and Francis Group, Boca Raton London New York.

[20] Mahmoud M.A.W., Ghazal M.G.M. Estimations from the exponentiated rayleigh distribution based on generalized Type-II hybrid censored data. *J Egypt Math Soc.* 2017;25(1):71–78.

[21] Merovci, Faton, and Ibrahim Elbatal. "Weibull-Rayleigh distribution: theory and applications." Appl. Math. Inf. Sci 9.5 (2015): 1-11.

[22] Mudholkar, Govind S., Deo Kumar Srivastava, and Marshall Freimer. "The exponentiated Weibull family: A reanalysis of the bus-motor-failure data." Technometrics 37.4 (1995): 436-445.

[23] Nadarajah, S. and Tahir, M.(2015). Parameter induction in continuous univariate distributions. Anais da Academia Brasileira de Ciências, 87(2), 539-568 .

[24] Oja, H., On Location, Scale, Skewness and kurtosis Of Univariate Distributions , Scandinavian journal of statistics 8(3) (1981), 154-168.

[25] Pal, Manisha, M. Masoom Ali, and Jungsoo Woo. "Exponentiated weibull distribution." Statistica 66.2 (2006): 139-147.

[26] Rényi, A . (1961). On measures of entropy and information. Proceeding of the Fourth Berkeley Symposium on Mathematics, Statistics and Probability 1, University of California Press, Berkeley, 547-561.

[27] Rayleigh L. Xii. on the resultant of a large number of vibrations of the same pitch and of arbitrary phase. *London Edinburgh Dublin Philos Mag J Sci.* 1880;10(60):73–78.

[28] Rinne, H., The Hazard Rate (Theory and Inference), (2014), Justus-Liebig University, Germany.

[29] Sabah Muhammad, and Kareema Abad Al-Kadim. " Odd Generalized Exponential Weibull Exponential Distribution." Journal of Engineering and Applied Sciences. 14. 8(2019).10360-10368.

[30] Saghir, A., Khadim, A. and Lin., Z., The Maxwell Length-biased distribution:Properties and estimation, Journal of Statistical Theory and Practice, 11(1) (2016) 26- 40.

[31] Sarhan, A., Ahmad, A. and Alasbahi, I. (2012c). Exponentiated generalized linear exponential distribution. Applied Mathematical Modelling. pp1-12.

[32] Tahir, M. H., et al. "A new Weibull-G family of distributions " Hacettepe Journal of Mathematics and Statistics 45.2 (2016): 629-647.

[33] Weisstein, Eric W. "Binomial theorem." https://mathworld. wolfram. com/ (2002).

Kolibri 360 GmbH
Ferdinand-Jühlke-Straße 7
99095 Erfurt, Deutschland
produktsicherheit@kolibri360.de

Druck:
CPI Druckdienstleistungen GmbH
im Auftrag der
Kolibri 360 GmbH
Ferdinand-Jühlke-Str. 7
99095 Erfurt